Neon Signs

Manufacture – Installation – Maintenance –

A History of the Neon Tube Light and the Electrical Sign Making Industry

By Samuel C. Miller

President of the Tube Light Engineering Company, New York.

And

Donald G. Fink

Associate Editor, Electronics

PANTIANOS
CLASSICS

Published by Pantianos Classics

ISBN-13: 978-1-78987-156-2

First published in 1935

Variety of Signs Visible in Times Square, New York.   (Courtesy of Brooklyn Edison Company, Inc.)

Contents

Preface

This book has been written with two types of reader in mind: the newcomer who desires to set up a neon-manufacturing plant, and the experienced man who is already set up in the business. The types of information required by these two types of men are different, of course. The newcomer needs to know how neon tubes operate, how to set up the plant equipment, and how to use it for manufacturing tubes. He will not be interested in the more advanced problems until he has mastered the fundamentals. The experienced man, on the other hand, has his equipment set up and in working order. He has manufactured tubes with it. He is interested in producing better tubes, increasing their life, improving his equipment for higher speeds of production, or solving such technical problems as obtaining unusual color effects and new types of animation.

In order to provide a useful book for both types of men, the authors have endeavored to present a complete picture of the manufacture of neon signs, their installation, and maintenance, beginning with the fundamental processes and operations and ending with the more advanced problems. The newcomer will find that he has need, or eventually will have need, of all the information presented. It is recommended, therefore, that he read the book through, as far as Chap. Fourteen, before undertaking to install equipment or manufacture signs.

The experienced neon manufacturer, on the other hand, will be familiar with many of the processes described. He will find the book most helpful as a reference guide, for use both in solving unusual problems and in mastering particular phases of the work that may give him trouble. For his benefit, the following outline of the book is presented.

The introductory chapter describes the history and present state of the neon industry. Following, the first section of the book describes the fundamentals of the neon sign: how it works, what materials and equipment go into its manufacture, and how these materials and equipment are related to the effectiveness and life of the sign. Although it might be supposed that these chapters would be an old story to the experienced manufacturer, the authors have found that even the most experienced men do not understand these fundamentals thoroughly, and that this lack of understanding is a frequent source of trouble. It is recommended that the experienced man read these pages carefully. If he finds nothing new, no harm is done; but in almost every case he will find some hint that will help him to produce better signs.

The second section of the book deals with shop practice, that is, it gives specific instructions for installing shop equipment and for using it in the proper manner. This section of the book will be largely a review for the experienced man, and he will need to consult it in most cases only for reference. Chapter Fourteen, however, is intended particularly for the old hand at the game. The two-color transformer chart, for example, is a new development that every sign maker can use to advantage, regardless of his experience.

Throughout the book the aim has been to be clear, complete, and practical-minded, and it is the hope of the authors that both the newcomer and the experienced manufacturer will find the book satisfactory in this respect.

The authors are indebted to the General Electric Company, the Acme Electric and Manufacturing Company, the Sangamo Electric Company, and the York Sign Company for illustrations used in this book. Valuable technical information has been supplied by the above companies and by the Anaconda Wire and Cable Company, the Underwriters' Laboratories, the Linde Air Products Company, the Air Reduction Sales Company, the American Gas Furnace Company, the Chicago Vacuum Equipment Company, the Beach Russ Company, the Corning Glass Works, the Eastman Kodak Company, the Central Scientific Company, and Signs of the Times. C. W. Homan, J. Comstock, and Keith Henney have rendered great assistance in reading the manuscript and in offering technical information. Miss Catherine Farrey, Mrs. Rosalie E. Miller, and Mrs. Dora M. Homan have likewise contributed greatly in typing and reading the manuscript.

Samuel C. Miller.
Donald G. Fink.
New York, N. Y., June, 1935.

Disclaimer

Chapter One - The Neon-Sign Industry - Introduction

In 1910, Georges Claude demonstrated his first commercial neon sign at the Grand Palais, in Paris. In the twenty-five years that have elapsed since that time, the neon-sign industry has grown in the United States until today it does an annual business conservatively estimated at thirty million dollars. From a single manufacturing plant in 1914, the number of plants engaged in neon-tube manufacture has grown until at least four thousand separate organizations are participating in the neon-sign business.

In the boom year of 1928, one neon company boasted a sales volume of nineteen million dollars, a tenfold increase over the record made only four years before. After a disastrous period of failures and retrenchment made necessary by the business collapse of 1929, the industry is now recovering and is growing steadily. No more convincing proof of this fact could be given than the records of the Sign Permit Bureau of New York. In 1933, 16,254 outdoor electric signs were installed in the Boroughs of Manhattan and Brooklyn alone. The sales price of these signs varies from $125 to $15,000 depending upon their size and complexity, but a fair average price is $200. At this rate, the signs installed in these two boroughs in a year were sold for a total of nearly four million dollars. In 1934, 19,550 signs were installed, an increase of more than 20 per cent over the previous year. No less than 90 per cent of all the signs installed were of the luminous-tube, or "neon," variety.

It is estimated that for every outdoor sign there are five neon signs installed indoors, at an average price of approximately thirty dollars. In all, nearly six million dollars' worth of neon signs are sold each year in Greater New York, and the business continues to grow.

That the neon business is a large one must be suspected by anyone who has lived in or traveled to any of our large cities. From one spot in Times Square, New York, more than 300 separate neon installations may be seen in operation at one time, ranging in size from a small window display to an animated roof-type structure covering 1500 square feet.

A good percentage of the business is done by large and well-established firms, but it is nevertheless true that there are many hundreds of small sign-making establishments sharing the business. The largest plants have invested thousands of dollars in equipment, but the small and often highly successful sign craftsman has equipment which cost him not more than five hundred dollars.

The neon-tube industry has had a unique product to sell. Just how popular neon signs have been with advertisers can be guessed from the fact that one concern in the business attained a million dollar volume of business sixty days after production was begun. But the history of the industry has not been free from strife and unpleasant dealings. Stock-selling promotions, years of wrangling over patent rights, heavy competition which until recently has been of cutthroat fierceness, and the dreary struggle of the independent manufacturers to find a berth in the industry have discouraged many able men from entering the business, and it is only recently that the industry has attained any degree of stability.

Why Is the Neon Sign So Popular? - Throughout the entire history of neon-tube enterprise, the public has shown remarkable interest in neon signs. From the point of view of the man in the street, the first neon signs had the fascination of novelty. But surely no one could claim that neon tubes are novel today. Long after the novelty wore off, the neon sign retained its sales effectiveness. There are many reasons why it has done so.

In the first place, the color of the light is easy on the eye but nevertheless extremely compelling. Single-color light like that of the neon or blue mercury tube has an eye appeal which white light cannot offer. The exact scientific reason for this fact has not yet been established, but the fact itself is well known. For the same reason that color on a printed page adds greatly to its attractiveness, the brilliant lines of color provided by the luminous tube have added greatly to the attractiveness of outdoor advertising.

A fact of little importance to the public but highly regarded by the advertiser is the high efficiency of neon signs. For illumination, neon is not so efficient as the incandescent bulb, but for visibility it far outshines the filament-type lamp. The neon tube puts out five times as much red light for a given amount of power as does the incandescent lamp, and it is this red light which attracts the eye and which penetrates great distances. The brilliance and sparkle with which neon signs stand out on a rainy night are too readily apparent to require proof. Neon has been recognized as the most efficient source of red light commercially available.

Other colors, such as blue, green, yellow, "gold," pinkish white, and others which can be obtained from luminous tubes, have also been in great demand.

Used in combination with the neon red, these colors can be combined to provide a wide range of color effects. In many instances, color designs of great artistry and beauty have been obtained. Architects have included in the plans for new buildings various provisions for permanent tube lighting to emphasize the features of the architecture at night.

Not the least important reason why luminous tubes are so well adapted to outdoor advertising is the flexibility which the glass tubing provides for creating difficult designs. Many advertisers want to feature their trademark. The trade-mark, reproduced faithfully in glass and even in the proper colors, can be provided by the well-equipped sign manufacturer. Special figures, scrollwork, and animated designs are all well within the scope of the neon craftsman; in fact, there is almost no end to the possibilities afforded by the tubular light. New applications, new forms, new mountings are being brought forward every day.

The neon sign has been subject to the often-heard criticism that it is not a beautiful light. Whether or not this is true, it is certain that forward-looking sign craftsmen are evolving new signs which can be accepted by the cities in which they are placed as real additions to the picture presented by the city as a whole, rather than as eyesores to be tolerated because they make money. The neon sign has been accepted since its invention as an outdoor advertising medium of unquestioned value. But with the growing use of color balance and graceful forms, these signs are also beginning to be looked upon as a distinct credit to the concerns who have installed and are operating them. This fact, while not considered by most sign makers today, will undoubtedly play a very important part in the future of the industry.

The Background of the Industry. - The first luminous tubes did not employ neon or any of the other rare gases. From 1893 to 1910, the so-called Moore tubes were prominent in the larger cities, but these were filled with nitrogen and carbon dioxide, two fairly common gases. For reasons explained in the next chapter, these tubes had a very short life.

In 1910, Georges Claude introduced the first commercial neon sign, and from that time until the outbreak of the World War, he was busy bringing his discoveries to the point where they could be introduced on a large scale. Throughout the war, little of importance developed, although during that time the fundamental Claude patent was granted, on January 19, 1915.

Up to the time of the use of neon for signs, incandescent bulbs, glass letters, and floodlighted painted signs were used for outdoor advertising. But unless a great many bulbs of high wattage were used, the signs were weak both in visibility and in penetrating power. Trouble from broken bulbs and burned-out filaments made such signs costly to maintain. Between the time a bulb burned out and the time it was replaced, the sign became definitely unattractive.

By 1925, a number of concerns had been licensed under the Claude patents and had begun operations. Neon tubes began rapidly to replace the incandescent lamp signs, and the public interest in the neon displays grew steadily. Before 1930, the public demand for neon signs had grown to such proportions that the words "Claude neon" had become a household phrase. In fact, many people believed that the man who started the business was a man by the name of Neon, a mistake often made even today. Within a short, time, all of the neon plants found themselves overtaxed with orders. This situation soon led to the formation of a host of companies rivaling the Claude organizations; at least seven other organizations were formed on a large scale, and many smaller ones also appeared. All of the larger companies rode on the tidal wave of the business boom from 1925 to 1929.

The huge sales totals and bright prospects at once led to commercial expansion. A great deal of stock was sold some of it on a legitimate basis, and a great deal in wildcat promotions. Within five years, it was estimated that 50 per cent of the bulb signs in the United States had been replaced by neon tubes. The business was growing to such an extent that the attempt to dominate the field gave rise to fierce commercial strife. One of the outstanding displays in this contest was the patent war which took place late in the 1920's and up to 1932. Early in the struggle a substantial number of the claims made by the Claude interests were declared invalid by the courts, while the fundamental claim on electrode size was declared valid. This litigation cost thousands of dollars in attorney's fees and in attempts to influence public opinion.

On January 19, 1932, the fundamental Claude patent covering the size of cold electrodes for neon tubes expired, and at once the patented electrode became available to the trade at large. Electrode manufacturing plants sprang up overnight; radio-tube machinery was put to use in neon plants, and the country was deluged with the once jealously guarded Claude electrode. In the Eastern and Midwestern parts of the country, decidedly unfavorable tendencies began to appear. The large concerns found more and more of their business going to the small independents. It was soon found that a man needed little more than a paintbrush, a layout table, and a low-rental loft to enter the field. A great influx of every type of man into the business was the result. Restaurateurs, jewelers, clothing manufacturers, and tradesmen of all sorts became interested in the neon-tube business and set up small plants.

Such conditions could lead only to cutthroat competition of the worst sort. Drastic inroads were made in prices; price competition went to the bitter limit. Expensive signs were sold on a small down payment to buyers of dubious credit. Guarantees of tube life were stretched to two and even three years. Some signs were installed with no down payment at all. Under these conditions, failures in the business were frequent, and the credit situation was in bad shape.

In many cases, neon-sign manufacturers had another line to fall back on; in many cases, profitable commercial-sign contracts provided the income for the support of the neon department. These companies maintained their good name and were a distinct credit to the industry during a very trying period.

With the repeal of prohibition, the industry received a definite advance. Old and new breweries and liquor distributors decided almost to a man on the neon sign as one of their major advertising mediums.

The future of the industry seems assured. The ever-increasing use of luminous tubing for straight sign advertising, and in many special applications as well, shows no sign of falling off. It is a business which can return a fair profit on a very small initial investment, a business in which success depends largely on the ability of the sign or electrical craftsman to master the processes involved in tubing manufacture and to market the signs on a fair basis. And it is becoming increasingly clear that the small plant, located in or near a medium-sized town, will supply the needs of that town over the competition offered by the more distant manufacturers. Transportation of neon tubes is costly; and it has been shown clearly that the most economical way of selling signs is to make them near the place of installation. It seems very probable, therefore, that each small town will have its local neon shop.

This book is designed to present the essential facts involved in the design, manufacture, installation, and maintenance of neon signs. As such, it is hoped that it may serve as a guide both for the man entering the business and also for the well-established craftsman who must occasionally solve unusual problems.

Part One – Fundamentals – The Why and How of Sign Making

Chapter Two – The Luminous Tube

The Luminous Tube: What It Is and How It Works. Fundamentally, the luminous tube is a very simple device. It consists of a glass tube, made specially vacuum-tight and fitted at each end with a metal terminal (or electrode). Inside the tube is a small amount of a rare gas. Connected to the two electrodes is a source of high-voltage electrical power. When the current is turned on, the tube glows with a steady piercing light.

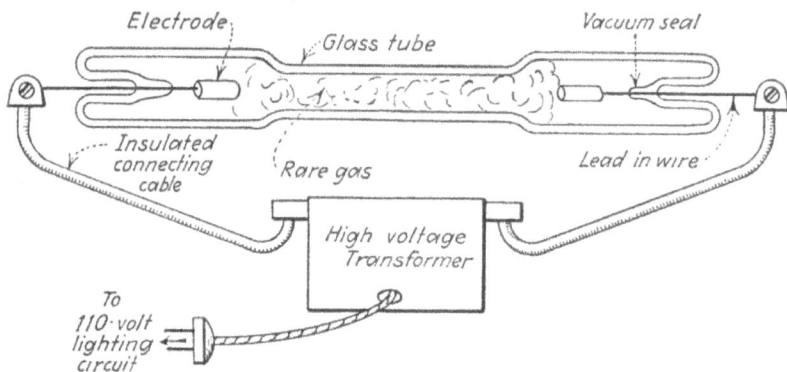

Fig. 1.—Schematic diagram of gas-filled tube and transformer.

The practical man can learn how to bend the glass, seal on the electrodes, pump out the air, fill the tube with neon or some other gas, seal it up, mount it, connect it to the transformer, and install the completed sign. He can do all of these things without knowing how or why the sign works. But if his newly installed sign goes dead in two weeks, and the customer demands satisfaction, that same practical man can save himself a great deal of trouble if he knows the simple facts about the glow discharge, what it is, and under what conditions it will deliver satisfactory service.

Sparks and Glow Discharges. - The familiar glow of the neon sign is a first cousin of man's oldest electrical acquaintance, lightning. The red glow of neon and the blue flash of lightning are both electrical discharges in gas. In the neon sign, the gas is neon, a rare gas of great value. In lightning, the gas is the air, a mixture of oxygen, nitrogen, carbon dioxide, slight traces of hydrogen, and still slighter traces of the rare gases. In the neon sign, the pressure is very low a partial vacuum, in fact in lightning, it is the normal atmospheric pressure. But the neon glow and the lightning flash are electrically very much alike. They both illustrate the fact that when electricity passes through a gas or a mixture of gases, light is given off.

If lightning and neon-glow are so much the same, what makes their light so different? Just the differences given above, that is, differences in pressure, and in the kind of gas used. The lightning spark is hot and blue-white in color. It consumes a tremendous amount of power. By using a special gas and a reduced pressure in the neon sign, it is possible to produce a steady glow, with very little power and with very little heat. The combination of pleasing, penetrating light, and high economy, with the flexibility provided by the glass tubing, makes the luminous tube a highly effective source of light for advertising purposes. The large industry based on luminous tubes, as outlined in the previous chapter, has resulted from the practical application of this low-pressure electrical gas discharge.

What Happens inside the Tube. - To understand what happens inside a neon tube, it must first be realized that the gas inside the tube consists of millions upon millions of particles of gas, called *molecules*. Under ordinary conditions, these molecules are electrically neutral; that is, they are neither positively nor negatively charged. To start the gas discharge, we must somehow break these molecules up into electric charges. If we do this, a current can flow through the gas, and this current flow will be accompanied by the desired light.

Each molecule is made up of negative *electrons* and positive *ions*. When the molecule is undisturbed, it has as many electrons as ions in it, and it is therefore electrically neutral. But if we remove a negative electron, the remainder of the molecule becomes positively charged; the result is the division of the gas into positive and negative charges.

The familiar law that like electrical charges repel one another and that unlike charges attract each other leads to the fact that a free electron is strongly attracted to any nearby positive ion. This attraction leads to the rapid merging of the positive and negative electrons into one neutral molecule. As they merge, *the molecule gives off light*. [1] This light is the light that we want to produce. Its color depends upon the kind of molecule, that is, upon the kind of gas.

Thus to produce a gas discharge we must continually remove electrons from neutral molecules and let them recombine with positive ions to form other neutral molecules. That is all that is necessary. If we could do that without the use of a high voltage transformer, we could have a neon sign that did not depend on electric current. But the only practical way of producing this ionization (as the electron-ion combination act is called) is passing a current through the gas. To pass this current through, we must have high-voltage electricity.

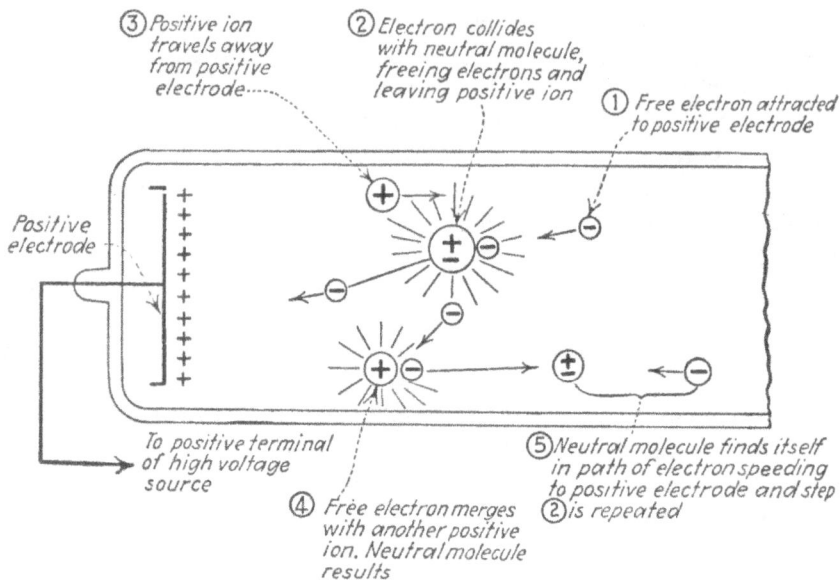

③ Positive ion travels away from positive electrode......

② Electron collides with neutral molecule, freeing electrons and leaving positive ion

① Free electron attracted to positive electrode

Positive electrode

To positive terminal of high voltage source

④ Free electron merges with another positive ion. Neutral molecule results

⑤ Neutral molecule finds itself in path of electron speeding to positive electrode and step ② is repeated

FIG. 2.—Action of electrons and positive ions in gas-filled tube.

How the Current Produces the Glow. - When the high voltage is applied to the electrodes, one electrode becomes, momentarily at least, positively charged, as shown in Fig. 2. It attracts, therefore, any free negative electrons which may be floating around in the gas. *If the voltage is high enough,* the electrons will be attracted with tremendous force and will speed toward the positive electrode with a speed of thousands of miles per second. Before one of these electrons can get very far, however, it collides with a neutral gas molecule which lies in its path. It hits this molecule with terrific force and liberates from it more (usually more than one) electrons which, once free, start off toward the positive electrode with great vigor. These electrons soon smash other neutral molecules, liberating still more electrons, which go off to smash still other molecules. In an unbelievably short time after the voltage is applied to the tube, the whole body of gas inside is in a frenzy of motion, electrons being liberated from molecules, free electrons combining with positive ions, giving off light as they do so, and then being blasted apart again.

The voltage supplied by the high-voltage transformer is *alternating;* that is, it reverses itself sixty (on some systems, twenty-five) times per second. Because of this fact, each electrode takes turns being the positive one, sixty times per second, and as a result the glow is distributed evenly over the whole tube. If direct current had been used, with one electrode remaining positive, the situation shown in Fig. 3A would occur. This undesirable condition is avoided by the use of alternating current.

14

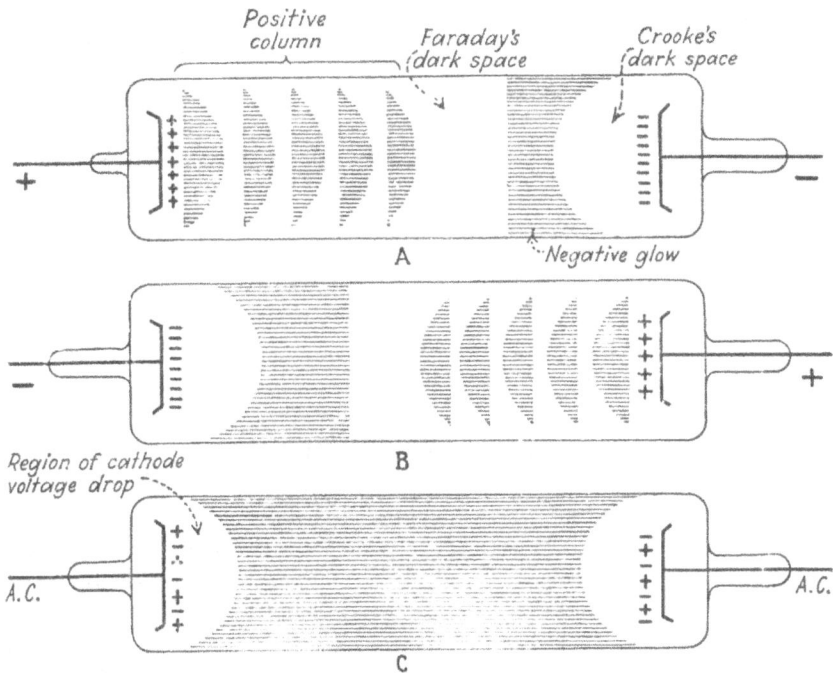

Positive column

Faraday's dark space

Crooke's dark space

+

−

A

Negative glow

−

+

+

B

Region of cathode voltage drop

A.C.

A.C.

C

FIG. 3.—*A* and *B* show type of discharge which results when direct current is used. When alternating current is applied to the tube, the even distribution shown in *C* results.

The Reason for High Voltage. - Unless high voltage is used, the electrons will not be attracted strongly enough to start the ionization process, no current will flow, no light will appear, and the sign will not operate. Hence, high voltage (as high as 15,000 volts) must be used. Furthermore, the voltage must be particularly high to start the sign's operating. Once the glow discharge has begun, however, less voltage is required to keep it operating. In fact, if the original high starting voltage were maintained, the current would be too great, and the gas would heat up excessively. The sign would consume a great deal of power, become very inefficient, and its life would be greatly decreased. So some means must be provided for lowering the voltage after the glow has started. A special type of transformer, made especially for the neon trade and called by its manufacturers a "high-leakage reactance transformer," performs this voltage reduction automatically. The method used will be described in detail in a later chapter.

Operating Current. - The electric current which passes through the luminous tube is an important quantity. If the current is too low, the sign will lack brilliance; if it is too high, the sign will overheat and have a very short life. The correct operating current depends upon the diameter of tubing, the kind of electrodes used, the kind of gas, and its pressure; each of these factors must be correct, so that the operating current is neither too small nor too large. The current of neon signs is measured in *milliamperes* (a milliampere

15

is one-thousandth of an *ampere,* which is the standard electrical unit of current). Operating currents in standard signs run from as low as 10 to as high as 50 milliamperes, and in some special lamps currents from 100 to 1000 milliamperes or more are used.

The current through the tube is accompanied by the passage of free electrons and ions from the gas into the electrodes. The heavier the current the more electrons or ions will flow to the electrodes in a given time. Each time that an ion hits the electrode, the energy with which it hits is transformed into heat, and as a result the electrode heats up. Hence, the heavier the current the hotter the electrode becomes. If the current is excessive, the electrode will begin to disintegrate, or even to melt. Hence, electrodes must be designed to withstand considerable heat, and even with proper design they cannot stand up if the operating current is too high.

The Effect of Gas Pressure. - We have already seen that a steady glow is possible only if the gas in the tube is at a reduced pressure. By "pressure" is meant the *force* with which the gas inside the tube presses against the glass wall of the tube. Pressure might be expressed in pounds per square inch, as air pressure is often expressed (normal air pressure, or "atmospheric pressure," is about 14.7 pounds per square inch). Usually, however, pressures in neon tubes are expressed in terms of the height of the column of mercury which that pressure will support. At sea level, air pressure will support a column of mercury to about 760 millimeters (29.9 inches), and the pressure of the atmosphere is thus referred to as a pressure of 760 millimeters. When the neon tube is pumped out, the small amount of gas which still remains in the tube can exert only a small pressure, compared with that of the air. The gas in a neon tube is usually at a pressure of 10 millimeters, or about one-seventy-sixth of atmospheric pressure. Vacuum pressures used in neon work often go down as far as 0.001 millimeter.

The amount of current which will flow through the tube depends largely upon the pressure. To realize why this is true we must remember that in a gas at low pressure there are fewer molecules per cubic inch than in a gas at higher pressure. That is, in the low-pressure gas, the molecules are fewer in number and farther apart from one another. In the low-pressure gas, therefore, the free electron has a longer distance in which to get up speed before it hits a neutral molecule, and, as a consequence, when it hits, it hits much harder, more free electrons are liberated than would otherwise be freed, and the whole action is much more intense. Hence, as the pressure is lowered, the current will increase. This is the general rule for pressures encountered in neon signs.

If, however, the pressure is reduced so low that a nearly perfect vacuum exists inside the tube, then there are so few molecules actually available that the current must decrease 1 from sheer lack of electrons and ions, even though the electron speeds are very high. Consequently it is found that after a certain point is reached in lowering the pressure, the current begins to decrease again. The effect of pressure on current is shown graphically in Fig. 4.

The range of pressures used in luminous tubes, between 3 and 20 millimeters, shows that in this range decreasing pressure means increasing current.

Summary. We now know that the current, and the resultant light in the tube, will increase as the voltage is increased and that it will increase if the pressure is lowered.

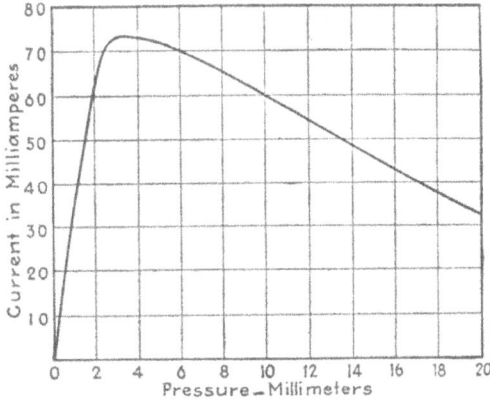

Fig. 4.—Curve showing increase of current as pressure is reduced (tube fed from constant-voltage source). Note decrease of current when pressure falls below 3 millimeters.

We also know that excessive currents are extremely harmful to the tube, since they will injure its electrodes and shorten its life. The moral is plain. To make a successful tube, the voltage and pressure must be *right.* They must be adapted to the gas or mixture used, and they must be maintained throughout the useful life of the tube.

The Effect of Diameter of the Tubing. - Neon tubes for sign-lighting purposes are rarely made with tubes larger than 15 millimeters (about 5/8 inch) or smaller than 7 millimeters (about ¼ inch) in outside diameter. These sizes are convenient to work with and thoroughly practical for almost all applications. It may be wondered why tubes larger than this are never used. The reason is that the current available from standard transformers is limited to about 30 milliamperes (some 60-milliampere transformers are used). This amount of current is capable of filling a 15-millimeter tube with light at the standard neon pressure. If it were used with a larger tube, the light would be spread so thinly throughout the tube that its effectiveness as a light source would be very poor. Practically speaking, therefore, the sizes of tubing are limited between the two values stated above. In general, the smaller the tube used the more brilliant the light. To obtain the maximum brilliance with the larger sizes of tubing, it is sometimes necessary to use slightly higher operating currents than those used with the smaller tubing. Figure 5 shows the relation between the necessary voltage and the diameter of the tube.

The Gases Used for Luminous Tubes. - No attempt will be made to go into details here concerning the gases used in luminous tubes, since they are fully described in the following chapter. In this introduction, only their elec-

17

trical characteristics will be discussed, in relation to the other electrical properties of the tubes. The rare gases neon, argon, helium, xenon, and krypton are ideally suited for use in signs, because the voltage per foot needed to produce a suitable brilliance of light is much less than that required for the more common gases, such as nitrogen and carbon dioxide, which also have been used for sign making. Stated in other terms, the resistance of the rare gases is among the lowest of all the gases available for the purpose. The gases with the lowest resistance are not those which produce the most light, however. Argon, for example, has a very low resistance, lower than that of neon, but its light is comparatively weak. To obtain the advantages of low resistance and good light emission, gases are often combined, low-resistance gases being mixed with the good light givers to produce a compromise having the advantages of each type of gas.

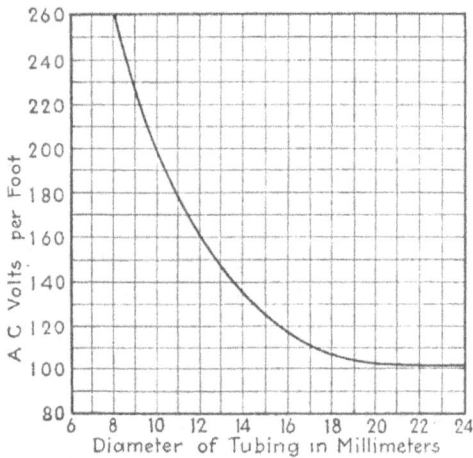

FIG. 5.—The voltage per foot required to operate tubing of various diameters filled with neon.

The Use of Mercury. - Strange to say, the most efficient and only practical way of producing a blue light does not involve the use of a gas but of mercury vapor. Vapors differ from gases in that a vapor cannot exist except as it evaporates from a liquid, while the gas can exist of its own accord. To obtain mercury vapor, therefore, it is necessary to insert liquid mercury in the tube.

To set a current flowing through mercury vapor is not so easy as through a gas, and for this reason a gas (usually argon or argon-neon mixture) is mixed with the vapor to aid the current flow. This extra gas also contributes somewhat to the light of the discharge, but its main function is that of a current carrier. The mercury vapor, like other vapors, is very much affected by changes in temperature; it will condense as the temperature is lowered. Thus, the light will become much more intense after the sign has had a chance to warm up. In extremely cold weather, the sign may never come up to full brilliancy unless it is properly constructed. All these factors must be taken into account in designing the sign for the particular job that it is in-

18

tended to fill. In general, the smaller the diameter of tubing used the higher the resistance per foot and the hotter the tube will get, allowing the mercury to vaporize more readily. In cold weather, however, small-diameter mercury tubes are often found to fade out while larger tubes remain bright, because of the greater heat conduction of the smaller tube.

Chemical Effects inside the Tube. - The action of the luminous tube in producing light of the required brilliance and color depends wholly upon the *electrical* ionization of the gas within it. If the action could be restricted to a purely electrical one, we should not have to worry about the sign's continuing to act as it should. Unfortunately, there are many possible *chemical* actions which may start within the tube if special precautions are not taken. And these chemical actions almost universally lower the efficiency of the tube. If any impurities, such as dirt, grease, or impure gas, are left in the tube after it is sealed off, these impurities, under the action of heat and electrical stress, will become chemically active. As they combine with one another, they may blacken the glass, they may combine with the metal of the electrodes, or, if the heat is intense, they may liberate gas inside the tube. If this last possibility occurs, the unwanted gas will become ionized, and it will give off light but not of the desired color and generally useless as a light source. This accounts for neon tubes' turning blue when they go bad. Many failures in sign making result from insufficient care in eliminating the impurities during pumping and filling. The gas inside the tube must be pure, and the glass and electrodes must be thoroughly clean if undesirable chemical effects are to be avoided.

Bombarding, an Essential Operation in Tube Making. - Removing the chemical impurities is not so easy a task as may be supposed. Mere washing of the tube and electrodes and the use of pure gas are not enough. Clean as the electrodes and gas may appear to be, they are actually harboring a great deal of gas and impurities which cling close to the surface and which cannot be removed by ordinary methods. The best way of getting rid of them is to heat the tube while it is being pumped out. The heat will drive the gas and other impurities from the surface of the metal and glass, in the form of some sort of gas or vapor. The pump will remove these gases, and if the heating process is kept up long enough all of them will be removed. The tube may then be filled with the rare gas and sealed off.

The most simple way of heating the tube and the electrodes is passing a current through the tube while it is still on the pump, when the air pressure has been sufficiently reduced to allow a heavy current to flow. This heavy current, usually much higher than the operating current, will pass through the low-pressure air in the tube, giving off much light and heat in the process. The light serves no purpose except to show that the bombardment, as it is called, is actually taking place. The heating has the desired effect of ridding the tube of its impurities. Bombardment is a most important and exacting process. An entire chapter (Chap. Nine) has been devoted, therefore, to explaining what it is and how it should be carried out.

Electrodes and Tube Life. Sputtering. - We have already learned these facts about electrodes: The electrode has the task of carrying current from the transformer wires to the gas. Since it is continually subjected to the bombarding of electrons and ions, it heats up, and it must be designed to withstand heat. Since the metal is hot, it is highly active chemically, and it may combine with gases or impurities within the tube. But by far the greatest difficulty with electrodes arises from what is known as *sputtering*. Sputtering occurs when the electrode, under the action of the heat to which it is subjected and the electrical forces which act on it, flies to pieces bit by bit. The metal of the electrode gradually flies off into space and coats itself on the inside of the glass tube. This effect in itself would cause no harm, since the blackening caused by the metal deposit is confined to the ends of the tubes near the electrodes. Eventually, of course, the entire electrode would be consumed by the process, but since the action is very slow, the electrode will nevertheless last for almost normal life. But sputtering is accompanied by a decrease of gas pressure in the tube. This loss of pressure eventually makes the tube inoperative.

The common explanation of sputtering action is as follows: As the metal electrode sputters to the glass walls, the metal particles trap the gas molecules and hold them between the metal coating and the glass wall. As a result, the pressure of the tube is greatly reduced. Finally the pressure is reduced so low that the sign loses brilliance, begins to flicker, and eventually goes dead. The gas in the tube is then said to have been "cleaned up" by the sputtering action. In practice, as the pressure falls off, the tube will operate at a very high temperature until the glass punctures at its weakest point, usually near the electrode. The tube is then entirely useless, and it must be taken down, opened up, cleaned, repumped, bombarded, and filled before it will glow as it should. In the early days, this sort of trouble was very common; in fact, the short life of tubes (due to sputtering) was one of the greatest hindrances to the commercial introduction of neon lighting.

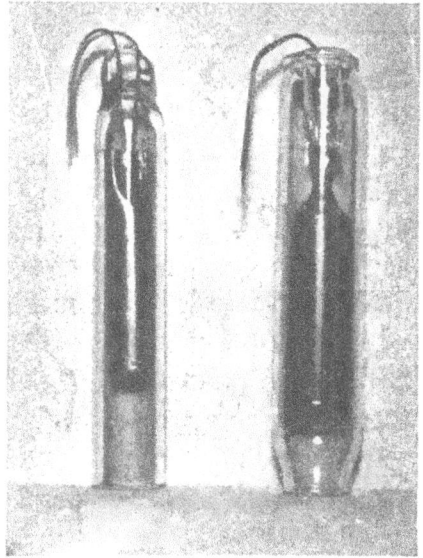

Fig. 6.—Neon-tube electrodes. Two types of construction are shown: at the left a pinch seal, at the right a ring seal.

Recently the theory of the "cleaning up" of gas due to sputtering, long held by neon men to be caused by trapping of the rare gas between the glass wall and the sputtered metal coating, has been challenged. Dr. Harvey Rentschler has proved that the gas, instead of being trapped against the wall of the tube, is absorbed by the metal of the electrode itself, after sputtering begins. But whether the gas is trapped against the wall of the tube or in the electrodes themselves, the result is the same. The gas pressure is reduced, the operating temperature rises, and the tube goes bad.

Sputter Control. - The logical solution to this problem is to build electrodes which can "take it," that is, electrodes which will not sputter under ordinary conditions. Some sputtering will always occur, and the life of the sign is thus always limited. But if the sputtering action is controlled, the life of the tube will be predictable, and maintenance guarantees and costs can be figured safely.

First, electrodes must have a large area exposed for the dissipation of heat. If an electrode runs cool, it will of course sputter much less violently than if it is hot. Secondly, the electrode must be made of the kind of metal which will best resist the sputtering action. Special metals and the common ones have all been tried; the best electrodes to use for each gas, each pressure, and each operating current are now known and have been more or less standardized by the industry.

The electrodes can be treated chemically to reduce sputtering before they are put in the tube; special patented processes are sometimes used for this purpose, and great virtues are claimed by each manufacturer for his particular "secret process." This treatment, while expensive, is sometimes absolutely necessary to insure reasonable life for a sign. Treated electrodes are especially necessary for helium (white or gold) tubes. One of the more common chemical treatments is known as *borating*. Dipping copper electrodes in molten borax gives them an inactive coat which greatly helps in reducing sputtering action. Special mechanical features are also incorporated in electrodes to reduce the tendency to sputter. All of these features are treated at length in the section on electrodes in Chap. Three. Suffice it now to realize that electrodes must be carefully chosen for the job if the sign is to have a long life.

Mechanical Requirements of the Tubing. - The maintenance of the proper gas pressure can be threatened by a more direct action than that of sputtering. If the glass envelope is not completely vacuum-tight, air will leak in, and the sign will soon go dead. To assure complete airtightness, the glass should first be mechanically strong and thus not subject to cracks or other breakage. This requires great care in glass blowing, since, if the glass is cooled too quickly, it will be left in a very brittle and fragile state. Particularly for signs installed out-of-doors, such strains within the glass will cause trouble, since the sign is then subjected to extremes in temperature.

The second requirement for a vacuum-tight tube is concerned with the lead-in wires which connect the transformer wires with the electrodes. Except for these lead wires, the whole wall of the tube is made of glass. The leads must necessarily be made of metal, to conduct the current. To make a good joint between glass and wire, the glass must "wet" the wire, that is, adhere to it firmly. Molten glass and wire do not readily cling to each other in this way unless the wire is copper or copper coated.

When wire and glass are heated, as when the wire is sealed on the glass or

21

during the operation of the sign, both glass and wire expand. If the wire tends to expand more than the glass, it will press against the glass with great force. If this force does not crack the glass at once, it will subject it to a great strain which will crack it at the least provocation. It is important, therefore, that the glass and the wire expand at the same rate as they are heated. If they do so, no internal stress will occur, and the joint will be strong. In practice, a special alloy wire ("dumet" wire) is used which has a coefficient of expansion very nearly equal to that of the glass over the entire temperature range to which it will be exposed in manufacture and service. The seal will remain vacuum-tight and serviceable if it is thus properly made.

Electrical Protection for the Sign Installation. - Simple as a luminous tube is, it must be realized at all times that it is potentially a very dangerous piece of apparatus. When properly constructed and installed, it presents less hazard than an ordinary lamp socket. But if proper precautions are not taken, the high voltage used is definitely dangerous. Since the current is limited to 30 or 60 milliamperes in most cases, the neon sign usually cannot produce a shock of sufficient force to kill a man. But the shock from a neon sign is powerful enough to overthrow a man, and the resulting fall may easily kill or cripple him. Furthermore, the possibility of the public's coming in contact with the high voltage is far too great a hazard to be allowed.

The precautions which must be taken involve the thorough insulation of every metallic part in the high voltage circuit so that it may not come in contact with anything else. Simply putting the wires and electrodes out of reach of the prying fingers of children and innocent bystanders is not enough, because the wires may come in contact with metal moldings, showcase fittings, etc., which will conduct the high voltage directly to the unsuspecting public. To avoid this, the industry has evolved, with the help of the Underwriters' Laboratories, a very complete set of insulating fittings and connecting wires which completely shield the high-voltage wiring not only from the public but from the weather as well. Protection from weather is essential in outdoor signs, of course.

The wires which lead from the transformer to the electrode leads are thoroughly coated with a high-resistance insulation capable of withstanding more than 15,000 volts. Special porcelain bushings are provided into which the entire end of the tube, with the electrode, fits. The connection of the high-voltage cable to the transformer is likewise protected by a special porcelain bushing. The entire high-voltage circuit is thus completely inclosed with insulation from start to finish. When so protected, it is actually far less dangerous than the exposed lamp socket in the home.

On no account should any sign be installed, *even for experimental purposes inside the shop,* without complete insulation of this type.

The Operations Involved in Making a Luminous Tube. - By way of summarizing the foregoing brief and necessarily incomplete introduction of the fundamentals of sign tubes, the following outline of the construction methods is presented.

Equipment Required. - A large workbench covered on top with a fireproof covering is required for laying out the patterns and for the actual work of glass bending. A supply of running water is desirable for cleaning tubes. A supply of gas, with a special booster, for raising its pressure if the gas-main pressure is low, and a supply of air likewise supplied from a blower are necessary. A variety of gas burners, or "fires," connected to the gas and air supply, is used in heating the glass tubing to the melting point, so that it can be bent to the desired shape.

A complete pumping system is needed, capable of drawing a good vacuum in the glass tube; a supply of

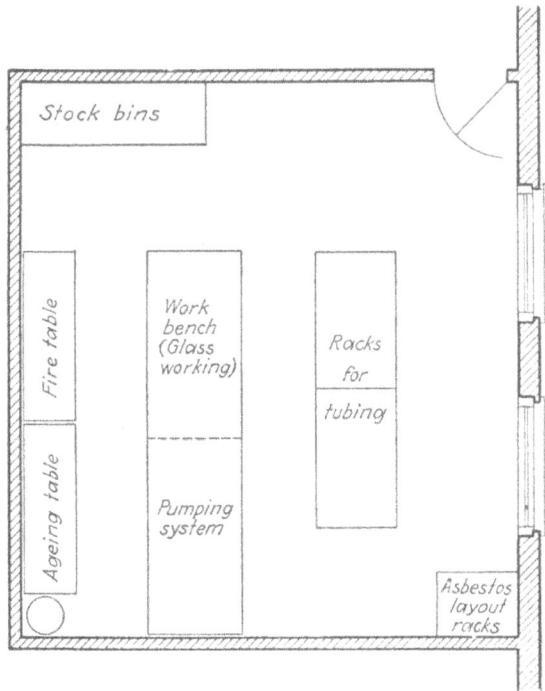

FIG. 7.—Typical neon-tube manufacturing plant layout.

rare gas is used in filling the tube after it has been pumped. A high-powered bombardment transformer must be available for heating the tube and electrodes during the bombardment process. Special tools for working the glass and for sealing the glass when the sign is completed are necessary. A testing coil for testing for air leaks, etc., is a great convenience. Finally, a supply of metal, porcelain, and glass parts for mounting the finished sign is necessary. For each sign two or more electrodes are necessary, and it is usual to have a complete supply of electrodes of several kinds, suited to particular jobs. If the sign is to be a mercury filled job, a supply of pure, clean mercury and some means of inserting it in the tube while the tube is on the pump must be provided. If all these parts, devices, and apparatus are available, work on the sign can begin.

Constructing the Sign. - The letters to appear on the sign are first laid out full size on drawing paper and then on asbestos paper, charcoal or carbon paper being used to produce a layout pattern for each letter and for the glass connections between letters, borders, etc. The glass is then heated, bent to shape, spliced together, and the continuous complete tube, open at each end, is formed. A hole is made in the tube, and a small tube or *tubulation* attached at a convenient point, for the purpose of connecting it to the pumps.

An electrode with its glass jacket is joined to each end of the tube. The tube is then tested roughly for leaks, by blowing through a rubber hose attached

23

to the tubulation. If tight, it is put on the pump, the tubulation being sealed to a corresponding size glass tube leading to the glass manifold. The glass stopcock attached to the blow hose is closed, and the stopcock to the pumps opened.

After the pressure in the tube has been reduced slightly, as indicated by the vacuum gauge, the stopcock leading to the pump is turned off so that the pump is disconnected from the tube. If the tube is now thoroughly airtight, the low pressure will be maintained.

The bombarding transformer is then connected to the two electrodes of the tube. The pump stopcock is opened, and the bombarder is thrown on. Just as soon as a discharge takes place, the pump stopcock is closed. The bombarder is then turned on again, the tube glows, a heavy current passes through, and considerable heat is generated. This bombardment is kept up for a length of time dependent upon the size and design of the tube and the particular conditions. The electrodes get red-hot. The glass eventually becomes so hot that it will scorch a piece of paper held against it. The bombarder is then turned off and disconnected.

The pump stopcock is then again opened, so that the pump again starts to reduce the pressure in the tube by exhausting impurities set free during the heat treatment.

The pump stopcock is then turned off, and another stopcock leading to the flask containing the rare gas is turned on slowly so that a small amount of rare gas is admitted to the tube. The proper pressure is measured with the vacuum gauge, and, when it has been reached, the rare-gas stopcock is turned off.

The proper sign-lighting voltage may now be applied to the tube from a regular transformer. The voltage disconnected, a tipping torch is applied to the tubulation leading to the pump, and the tube is melted off, or "sealed off," the vacuum system.

The tube is now complete except for the process of "aging," which is often necessary to bring it up to its proper brilliance. Aging is accomplished by connecting the completed tube to a transformer which will supply either the rated current or a current somewhat higher than rated. After a short period (not more than 15 minutes under ordinary conditions), the aging should be complete, and the sign will then operate at full brilliance on a transformer of the type which will be used in the installation. When the pumping system is in good working order and the tube is properly pumped before filling, no aging should be necessary.

Mounting the sign is a process which depends largely upon the particular installation. Usually, however, a background of metal is provided against which the tube glows, and which serves to emphasize each letter. This background may be made, as it often is, in the form of a box or housing in which the transformer can be mounted. On the background are mounted standoff brackets which seize the glass tube at several points and hold it firmly in place. At the electrodes, two special recessed bushings are provided which take the lead wires. From the bushings runs the high-tension cable which connects the lead-in wires to the transformer. This cable is connected to the transformer and insulated at terminals. The complete sign is shown in Fig. 8.

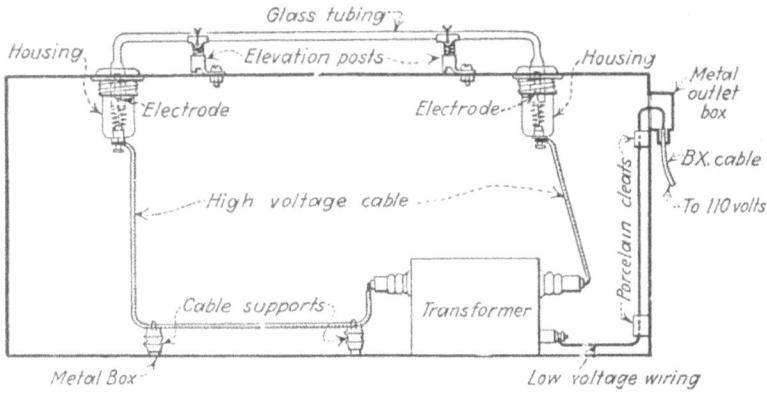

FIG. 8.—A typical box-type neon sign, showing proper location of transformer, mounting parts, cable, tubing, etc.

The transformer is supplied with current through a low-voltage winding, known as the primary winding, which is connected to the 110-volt lighting circuit. The usual practice of using armored BX cable for house wiring is generally followed in the low-voltage connections. A switch for turning the sign on and off is connected in this low-voltage circuit, and if a flasher, an automatic device for turning the sign (or its separate parts) on and off, is installed, this may be connected in the low-voltage circuit also, although flashers are often used on the high-voltage side of the transformer.

Care is taken in hanging the sign so that it will not fall while in service, and it should be protected from the possibility of being broken by near-by objects.

The sign, once installed, should give service (depending upon the care with which it has been designed and constructed) from one year to three years of continuous or intermittent action. At the end of that period, sputtering and other forms of decomposition will have advanced to a degree where replacement is necessary. The tubing may then be taken down, opened, cleaned, provided with new electrodes, repumped, rebombarded, and refilled.

The Maintenance Problem. - The limited life of even the best-constructed sign should not be forgotten by the sign maker who intends to make money in his business. He must know how long his signs can be expected to last under the various conditions to which they are likely to be subjected, and he must sell the sign in such a way that the customer understands the necessity for occasional replacement. Prices must be figured to take care of replacement, or, if this is not done, the cost of replacement must be assumed by the customer. Not the easiest problem in the neon-sign business is that of figuring life guarantees and then living up to them. Costs will inevitably mount if

25

the sign maker is continually servicing the signs that he has sold; therefore he must know how to build a long-lasting sign. If it does not come up to expectations, he must be able to analyze the trouble and repair it quickly and in such a manner that the trouble will not recur.

The Road to Successful Sign Making. - A bird's-eye view of the art of making and servicing luminous tubes having been given in this chapter, it now remains to go into the details of each part, its method of manufacture and use, and the various processes which are used to produce rare gas tubes. The foregoing introduction will serve to show how each operation, as it is described piecemeal in the following pages, fits in with the rest of the art.

The foregoing pages have been intended also to give the prospective sign craftsman some idea of the simplicity of the art. The neon sign is not a complicated structure, and it can be built by anyone who has the mechanical ability and intelligence to apply his knowledge. To build up a successful business is not so simple as building successful tubes, however. The best sign maker in the world must have a good business sense and be alert to his opportunities if he is to continue to get orders; otherwise, his competitors will take away his business in short order. But the foundation of a good neon-tube business is good solid craftmanship.

[1] This long-accepted theory has recently been challenged by J. J. Thomson, the famous British authority on gas discharges. He is inclined to think that the light is caused not by the recombination of electrons and ions but by the excitation of the ions themselves as they collide with one another. See J. J. Thomson, "Conduction of Electricity through Gases," 3d ed., p. 338.

Chapter Three - Materials Used in Constructing Tubes

The Rare Gases, Electrodes, Glass Tubing

The Atmosphere. - It was not until the middle of the seventeenth century that an English scientist by the name of Robert Boyle first attempted to analyze the true nature of the air. Since that time, we have come to know that the air we breathe is made up of many different gases, some of them very plentiful, and others extremely rare. Pure air, from which all the water, carbon dioxide, and dust have been removed, contains seven gases. The proportions of each are given in the table:

TABLE I.—GASES FOUND IN THE ATMOSPHERE

Gas	Amount by Volume
Nitrogen	1 part in 1.281 (78.060 per cent)
Oxygen	1 part in 4.760 (21.008 per cent)
Argon	1 part in 106.8 (0.930 per cent)
Neon	1 part in 80,800 (0.001 per cent)
Helium	1 part in 245,300
Krypton	1 part in 20,000,000
Xenon	1 part in 70,000,000

The last five in this list, argon, helium, neon, krypton, and xenon, are rightly called the *rare gases*. Just how rare xenon is may be realized from the fact that there is actually more gold dissolved in sea water than there is xenon in the air.

This interesting group of gases form a chemical family; that is, they display one definite chemical characteristic in common. All of these gases are chemically *inert;* that is, they will not combine with any other substance; they are always found free, never in combination with other substances. Chemists have never succeeded in making them combine either with each other or with any other chemical substance. This fact makes the rare gases ideally suited for use in luminous tubes. In the preceding chapter, we noted the injurious effects which may take place if the gas combines chemically with the electrodes or with impurities in the tube. The rare gases, because of their inert nature, cannot so combine.

Early Attempts to Use the Commoner Gases. - It was natural that the more common gases, such as nitrogen and carbon dioxide, should be tried first in the early attempts to produce successful luminous tubes. Although such gases are capable of producing a brilliant light, the tubes which used them had a very short life. The main reason for this fact was the chemical activity of the gases used; the nitrogen or carbon dioxide would soon combine with the electrodes. As this chemical combination went on, the pressure became lower and lower, until eventually all the gas was "cleaned up," and the tube failed to function.

In 1893, D. McFarlan Moore, a resident of Newark, New Jersey, began experimenting with luminous tubes using common gases. Soon convinced that they could never become commercially successful unless the problem of short life were solved, he hit upon the idea of using an electromagnetic valve for admitting more carbon dioxide or nitrogen to the tube as the current through the sign changed owing to gas cleanup. This invention, while cumbersome, was instrumental in bringing the tube-sign industry to life, and Moore tubes, as they were called, were a familiar sight during the early 1900's.

But the large quantity of gas required made it necessary to use tubes at least two inches in diameter and of great length. The signs were thus limited in application and so expensive that only the larger business houses could afford to install and operate them. Despite these drawbacks, Moore enjoyed a large business for ten years after his invention of the gas valve. One of the first installations was that made in 1904 for the newspaper *The New York World*, the words THE WORLD being formed in script from a single unit of large tubing. This sign was followed by many installations of a similar character, both in America and abroad.

The Discovery of the Rare Gases. - The rare gases were discovered in the following order: Helium was found to exist in the sun by spectroscopic analysis, in 1868; it was discovered on the earth some seventeen years later. In

1893, argon was discovered by Lord Rayleigh and Sir William Ramsay. In 1898, suspecting still other gases, Ramsay began a definite search for them and succeeded in isolating neon for the first time. A few days later, following the same track, he discovered krypton and xenon. The whole series of rare gases was thus revealed to science between 1880 and 1900.

The discovery of the rare gases at once led to their use in scientific laboratories, in an endeavor to learn their properties. But the extreme rarity of the gases and the consequent expense of extracting them from the air made any commercial use an economic impossibility. It was not until 1907 that Georges Claude, the famous French inventor, perfected a process for obtaining the rare gases cheaply. Working on processes which he and a German, Linde, had developed independently for the making of liquid air, Claude succeeded in laying the foundation of the neon-tube business.

Having workable amounts of rare gas at his disposal, Claude at once began the development of the rare-gas luminous tube, intending to introduce it commercially. His remarkable achievements in this field began in 1910, when he filled a nitrogen Moore-tube display in Paris with neon. Since that time, his business has expanded until one of the largest luminous-tube companies in the world now bears his name.

Commercial Methods of Extracting the Rare Gases. - The process which Linde and Claude developed for the commercial extraction of the rare gases depends upon the liquefication of air. Air, cooled by the usual refrigeration means, compressed by air compressors, and then allowed to expand suddenly through a small opening, can be made to lose heat. If the process is repeated, each time using the cool gas which results from the preceding expansion, the gas can be made successively cooler and cooler until eventually it becomes liquid. Liquid air, at a temperature of about -196 degrees centigrade (-371 degrees Fahrenheit), contains liquid nitrogen, oxygen, and liquid argon. These three liquids have three different boiling points. By allowing the nitrogen to boil off (since it has a lower boiling point than the argon), the liquid argon and oxygen are left. When the argon is allowed to boil, practically pure argon gas is the result. The same general process is applied for the separation of neon, helium, krypton, and xenon. Helium can be obtained much more cheaply from natural-gas deposits, with which it is mixed in considerably greater amounts than in the air, but when so obtained it is not so pure as when extracted directly from the air.

Neon Gas. - Neon, while not the commonest rare gas and while not discovered first, is by far the most widely used gas in the luminous-tube industry. It is colorless and odorless, slightly heavier than oxygen, twenty times as heavy as hydrogen, and five times as heavy as helium. It has an ionization potential (a measure of the voltage required to light it) lower than that of helium but higher than the other rare gases, argon, krypton, and xenon. Since the gas is not particularly well suited from either the standpoint of cost or electrical characteristics, we may well wonder why it has attained such universal popularity. The answers are its color, which at once caught the public fancy, and its high efficiency, which makes it one of the best sources of light for advertising purposes. The penetrating power of its red light, which shows up to best advantage in the poorest weather, has been utilized in beacons for aviation and marine service.

In an effort to establish the reason why neon lighting has this penetrating power, studies have been made of the light that it gives off. It has been found to be predominantly orange and red in color, but there are also small amounts of yellow and even green light. The relative amount of light of these different colors given off is shown in Fig. 9. The peaks in the curve show that the great-est amount of light is given off in the red and orange regions.

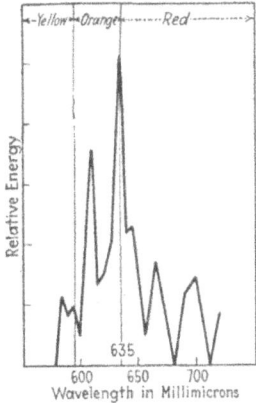

Fig. 9.—Relative light energy from a neon-filled tube. The light energy is confined primarily to the red and orange regions.

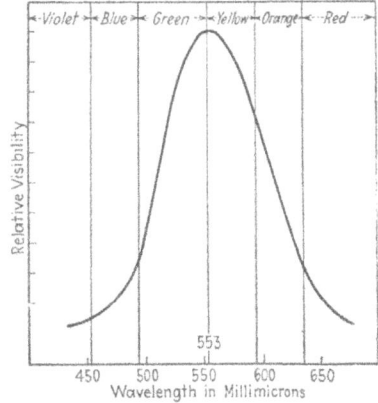

Fig. 10.—Relative visibility of different colors to the eye. Maximum visibility occurs in the green-yellow region.

Studies have also been made of the kind of light to which the eye is most sensitive. Figure 10 shows the relative visibility of the various colors in the rainbow. It will be seen that the maximum visibility is in the green-yellow region, while the orange-red region has much less effect. Why is it, then, that the neon glow has such a powerful effect? The answer is shown in Fig. 11, on which is plotted the ease with which different colors are transmitted through the air. The relative transmission varies according to the kind of weather, as shown by the three curves. For rainy weather, the maximum transmission occurs at a wave length of 635 millimicrons, the exact wave-length at which neon has its greatest output. The amazing fact is that a neon sign is 20 per cent more pene-trating during rain than dur-ing clear weather. This is the technical basis for the high penetrat-ing power of neon and one of

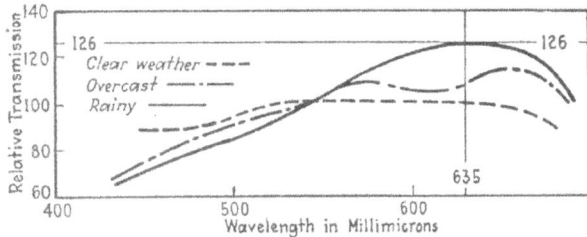

Fig. 11.—Relative transmission of different colors under various weather conditions. At 635 millimicrons, the wave length of neon light, the transmission is highest in the poorest weather.

the most important reasons for its high popularity.

The Light Efficiency of Neon Gas. - Since the neon tube is sold in competition, either directly or indirectly, with the incandescent-light sign, the comparative cost of operating the two types of signs is a most important fact. Table II shows the light efficiency, in lumens of light per watt of power, for various gases and diameters of tubing. In comparison with these values, a standard 10-watt Mazda lamp gives 100 lumens of light, or approximately 10 lumens per watt. To get the same illumination value from a neon tube as that provided by a 150-watt lamp, 20 to 25 feet of 11-millimeter tubing would be required.

TABLE II.—LIGHT OUTPUT OF LUMINOUS TUBES

Color	Rare gas used	Glass	Outside diameter, millimeters	Current, amperes	Lumens per foot	Watts per foot	Efficiency, lumens per watt
Red.....	Neon	Clear	11	0.025	70	5.7	12.2
			15	0.025	36	4	9.
Blue....	No. 50 or B-10*	Clear	11	0.025	36	4.6	7.8
			15	0.025	18	3.8	4.7
Green....	No. 50 or B-10*	Soft canary	11	0.025	20	4.6	4.3
			15	0.025	8	3 8	2.1

* Mercury used in blue and green tubes.

TABLE III.—READABILITY CHART
Neon (red) Tubing

Height of letter, inches	Distance visible, feet	Height of letter, feet	Distance visible, feet
2	65	3	1500
3	100	3½	1750
4	150	4	2000
6	200	4½	2250
8	350	5	2500
9	400	6	3000
10	450	7	3500
12	525	8	4000
15	630	9	4500
18	750	10	5000
24	1000	11	5500
30	1250	12	6000

The above values are approximate figures subject to change according to such conditions as current density, background, visibility (rain, fog, smoke), and the presence of other lights in the neighborhood.
For blue tubing, deduct 25 per cent; for green, deduct 35 per cent.

In sign work, however, the chief requirement is not illumination but visibility. Since red rays have great penetrating power, [1] they are the most important from the standpoint of visibility. A neon tube will provide 10 lumens per watt in the red range. The Mazda lamp transmits only 2 lumens per watt of the red. Hence, for the red ray the neon sign is five times as efficient as the Mazda lamp. This accounts for the generally accepted fact that to produce equivalent penetrating effect, an incandescent-lamp sign must be provided with two or three times as much power as the neon-tube sign requires for the same effect. [2]

Neon gas is sold in glass containers, each containing 1 liter (61 cubic inches, or about the same volume as that contained by a quart milk bottle). In a new container, the gas is at atmospheric pressure, that is, the same as the air on the outside, so there is no pressure on the walls of the container either inside or out.

The gas commercially sold is "spectroscopically pure" (less than 1 per cent impurities). It costs between eight and ten dollars per liter.

The gas inside the flask, while sufficient for many hundreds of feet of tubing, weighs very little. In fact, neon gas costs about $500 per ounce, many times the value of an equal weight of gold. The name neon comes from the Greek word *neon*, meaning "something new" and was given by Ramsay to the gas to signify its unexpected discovery. [3]

Helium. - Next to hydrogen, helium is the lightest gas known. For this reason, it is extensively used in the aircraft industry as a supporting gas for dirigibles. It is found mixed with natural gas and contained in certain minerals and exists in the atmosphere. It derives its name from *helios,* the Greek word for "the sun," and was so named because its existence was first shown in the sun by Janssen, who discovered a bright yellow line in the spectrum of the sunlight. Later, helium was discovered on earth and identified as the same substance present in the sun.

In the luminous-tube industry, helium is used to produce a yellow or whitish light. The gas has a high ionization potential compared with neon and the other rare gases; that is, it takes a higher voltage to produce the required brilliance of light. Because of their higher resistance, helium tubes take more power to operate than do neonfilled tubes and give off considerable heat. This condition gives rise to many problems and has prevented the wide use of helium tubing until recent years. The gas must be used at low pressure (approximately 3 millimeters) to give good brilliancy; red or blue signs made with such a low pressure would last less than 100 hours. The resistance per linear foot of the gas is about twice that of neon, and accordingly only half as many feet of helium tubing can be operated from a transformer as could be used if the tubing were neon-filled. For brilliant helium tubes, a 60-milliampere transformer is usually used.

The successful use of helium depends almost entirely upon the electrodes used. Special electrodes have been designed with chemical treatment intended to reduce sputtering to a low value. Practically all helium electrodes have a special chemical coat and require protection from the air. The electrode is sold, therefore, in a vacuum sealed jacket. A more detailed account of these electrodes is given in the section on electrodes. Helium costs from eight to ten dollars per liter.

Argon. - Argon gas, from *argon*, "the inactive one," exists to the extent of 1 part in every 106 of the air we breathe. Suspected as early as 1796, by Cavendish, its existence was not definitely proved until 1893 by Ramsay, who named it " argon" because of its extremely inert character. The gas is about twice as heavy as neon. In 1914, its commercial extraction was undertaken, not for the luminous-sign trade but because it was found that incandescent lamps filled with argon had a much higher efficiency and longer life than vacuum lamps. About one-half of all the Mazda bulbs manufactured today contain argon.

Argon ionizes with a blue color, but its light efficiency is not great, so that it is rarely used as a light source in gas-filled tubes. On the other hand, its extremely low ionization potential makes it an excellent conductor of electricity. It is used almost exclusively in the signtube art mixed with neon in a gas mixture containing approximately 80 per cent argon and 20 per cent neon. This gas mixture is used as the "carrier" gas in mercury filled signs. This carrier action is necessary because mercury vapor within a sign tube will condense at low temperatures; when this happens, the tube will fail to start or else give a very faint light. The argon-neon mixture has the unusual property of "supporting" the mercury vapor and keeping it active throughout a wide range of temperatures. The argon, having a low resistance, permits current to pass easily, and its blue ionization color blends with the blue of the mercury-vapor glow to give the desired color. The same gas mixture and mercury, when used with a yellow glass, produces the familiar green tube. The red, yellow, blue, and green signs thus depend on four elements: neon, helium, argon, and mercury vapor.

The argon-neon mixture is sold in the same sort of container as is used for neon and helium, and it retails for about the same price.

Krypton and Xenon. - Krypton, "the hidden" and xenon, "the stranger" (pronounced krip-ton and zē'-non), are so rare that their commercial use in signs has only just begun. Xenon, the rarest, is four thousand times as costly as gold, being quoted as $78,000 per troy ounce. Its spectrum color is sky-blue. Krypton, above 1 millimeter pressure, ionizes with a white color, tinged with lavender. These two gases have the lowest electrical resistance of all the gases in the rare group. Xenon is the heaviest of the group, being 6.5 times as heavy as neon. Pure krypton sells for 30 cents per cubic centimeter, that is, $300 per liter. Xenon costs 75 cents per cubic centimeter, or $750 per liter. These gases are almost always sold in mixtures with the other rare gases. When mixed with neon, argon, or helium, the cost per liter is the cost of the first gas, plus approximately $3 for each 1 per cent of krypton. No quotation is given for xenon, but it is approximately two to three times as expensive as krypton.

Mercury Vapor. - The first luminous tube ever constructed contained mercury. A simple evacuated tube containing mercury was found as early as two hundred years ago to give a faint greenish-blue light when shaken back and forth. The static electricity generated by the friction of the mercury against the glass walls of the tube ionized the vapor, and light was produced. Today, mercury vapor is universally used for blue and green luminous tubes. At ordinary temperatures, there is no such thing as mercury *gas*; ionization of mercury at room temperature can take place only if a voltage is applied to a body of mercury *vapor,* which cannot exist unless there is liquid mercury present. Hence, blue and green signs contain liquid mercury, which vaporizes under the proper conditions, thus providing mercury vapor.

If mercury vapor alone were used, the operation of the tube would be very erratic, owing to the effect of temperature on the pressure of the mercury vapor and owing also to the fact that the pressure of mercury vapor at ordinary room temperature is not sufficient to give a brilliant light. Hence rare-gas mixtures containing argon and neon or argon, neon, and helium are used to "support" the mercury vapor. If argon were used alone for this purpose, the tube would probably run too cool (since the resistance of argon is low). The neon thus serves the definite purpose of producing a hot tube. The heat of the tube vaporizes the mercury to the proper point.

Mercury signs usually take some time to start; that is, the argon-neon mixture ionizes first, producing heat, which vaporizes the mercury. Until the mercury has vaporized, it will not contribute to the glow, and as a result a short time is required for the sign to heat up and come to full brilliancy after it has been turned on.

Helium is sometimes used with the neon and argon in the carrier gas, because its high resistance adds still more to the heating effect. For mercury signs to be installed in cold climates or under other adverse conditions, this treatment is often necessary to secure satisfactory operation. The high-resistance helium-argon-neon carrier gas has the disadvantage, of course, of requiring a high voltage per foot of tubing. This reduces the number of feet which can be run from a standard 15,000-volt transformer to about 20 feet of 12-millimeter tubing.

At one time, xenon and krypton were considered as substitutes for argon in gas mixtures because of their low resistance. But these gases had no apparent advantage over argon-neon and because of their very great cost have not been used to any great extent. Krypton is used in hot-cathode window-lighting units, since the low ionizing potential aids in starting the tube.

A mixture containing krypton was developed also which would give a blue color without the use of mercury, but this mixture was found to be inferior to the standard mercury-argon-neon method.

Chemical Effects in Mercury Tubes. - Mercury in tubes has a very great disadvantage which must be carefully considered in the design and construction of green and blue tubes. Unlike the rare gases, it is very active chemically, especially when vaporized. It will combine with the electrodes or with any impurities in the tube unless the electrodes are specially protected and unless there is absolutely no foreign matter in the tube. The products formed when mercury combines with impurities are almost always black in color, and they will blacken the tube in short order.

Special care in cleaning the tubes with acid, a double bombardment, and the use of absolutely pure triple-distilled mercury are essential to the successful manufacture of green and blue tubing. Aging before the mercury is inserted is also recommended. These processes are all described in detail in later chapters.

Electrodes

Structure of the Cold Electrode. - Two types of electrodes are in use in luminous tubes: the cold electrode and the hot-cathode type. The electrodes described thus far have been of the cold type. The hot-cathode type of electrode is beginning to find use in neon-sign practice (a short description of it is given later this chapter). The hot cathode contains a filament or filament and metal shield attached together, which is heated by connection to a separate electric circuit; the heat provides a supply of electrons which allows the sign to be operated at lower voltages than the cold electrode requires.

The cold electrode must be operated at high voltages but because of its convenience compared with the hot-cathode type is used almost exclusively by the neon trade. The cold electrode has been the subject of a great deal of investigation by luminous-tube engineers ever since the first rare-gas sign was installed, and as a result much has been learned about the proper design of this part of the tube.

Because of the difficulty of properly constructing electrodes it is almost universal practice for the trade to buy the electrodes ready-made from supply houses. The electrodes supplied are of standard design and contain the following parts:

1. The electrode shell, which supplies the current to the gas in the tube.

2. The lead-in wires, which connect the electrode shell through the glass seal to the high-voltage cable and the transformer.

3. A glass jacket and seal, which surrounds the lead-in wires with a vacuum seal and which incloses the electrode shell in a cylinder of

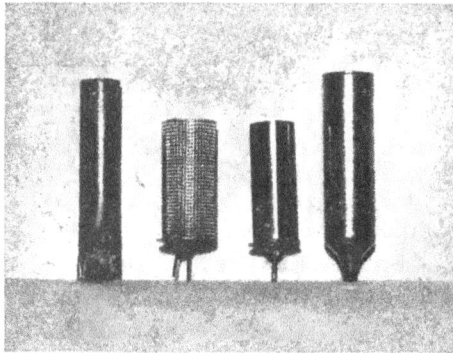

Fig. 12.—Electrode shells. The second from the left is of the "pertruded" form; the outer specimens are borated copper, and the second from the right is an iron type.

glass, open at one end. This jacket is joined to the end of the glass letter.

4. Some form of heat-insulating material which prevents the electrode shell from coming in contact with the glass jacket. In some cases, sufficient spacing between shell and glass is used, in place of other insulation.

These parts are shown in Figs. 6, 12, and 13.

It cannot be too strongly emphasized that success in building neon and other luminous tubes depends upon the use of properly designed electrodes. Hence, although the sign craftsman will not have to make them, he will profit if he knows how they are made.

Electrode Requirements. - The electrode shell, being an electric terminal, must be made of metal. Since the shell is inside the tube, and since it is subject to heavy electrical stress, it must be made of very pure metal. If the metal contains impurities, these impurities will be freed by heat and will form a gas. This gas, mixed with the rare gas in the tube, will greatly detract from the sign's usefulness efficiency. In fact, if the impurities in the electrode are excessive, they will cause the eventual breakdown of the tube.

Since the metal must be pure, alloys such as brass, bronze, or steel are generally unsuited to electrode use. Even though the brass or steel may be chemically pure, actually it contains much gas inclosed between its pores, which becomes free upon heating.

The metals used are electrolytic copper, pure nickel, purified iron (Swedish iron, known as Svea metal, is almost universally used for iron electrodes), and pure aluminum. If these metals are specially heat-treated before being manufactured, and if properly protected before use, they will give very satisfactory performance.

The second requirement for a good electrode is sufficient area for heat dissipation and for the purpose of carrying the tube current without sputtering. This requirement was the basis of the fundamental patent issued to Georges Claude (see page 49). Since the electrode gets hot during bombardment, it must be able to dissipate this heat without melting. It must also be large enough to withstand the electric forces which act on the electrode causing it to disintegrate slowly or sputter.

The degree of sputtering depends to a large extent upon what is known as the *cathode-voltage drop,* or, more simply, cathode *fall.* When the electrode is negative, as it is sixty times a second during operation, the positive ions are attracted to it. These ions, being much heavier than electrons, bombard the surface of the electrode with great force and cause it to wear away slowly. The force with which these positive ions act depends upon the drop in voltage at the electrode. The higher the cathode fall the greater will be ihe bombardment and sputtering. To keep the cathode drop low, two methods are available. The first and simplest is to make the electrode large in surface area. The second, more complicated but very widely used, is to coat the electrode chemically with a coating which reduces the cathode fall by liberating free electrons.

A third requirement for electrodes used in mercury tubing is chemical inactivity. When used with the chemically active mercury of the blue or green tube, electrodes may be injured by chemical action.

Lead-in Wires. - The lead-in wires must be able to carry the sign current without heating, and they must have the same coefficient of expansion as the glass in which they are inclosed. To meet these requirements a special wire, known as *dumet wire,* has been developed. This is an alloy of nickel and iron (about 40 per cent nickel), which is covered with a liberal coating of copper, the copper cross section being as much as one-fifth the cross section of the wire. This wire has ample current-carrying capacity (a 6-mil wire will carry 200 milliamperes); it has a coefficient of expansion closely matching that of lead glass, and the copper coat makes it adhere to the molten glass which is pressed around it.

The seal between the wire and glass must be completely vacuum-tight. The use of dumet wire practically insures this, but care must be taken in manufacturing the seal to allow the glass to cool slowly; otherwise it will contain internal strains which will crack the seal.

Electrode Insulation. - The electrode shell is kept from coming into contact with the glass jacket around it by the use of a ring of special heat-insulating and electric insulating material. Mica, porcelain, "lava," glass, or a special coating for the shell are among the materials used. Their purpose is to keep the hot electrode from touching the glass during the bombarding process. If it did so, it would immediately crack the glass. Some manufacturers rely on very strong and stiff lead-in wires to keep the shell centered within the glass jacket, and the method is a very good one, since the vacuum around the electrode is the best kind of insulator. The various forms and shapes which electrode insulation may take are shown in Fig. 13.

The glass jacket which forms part of the envelope of the completed sign must be carefully designed also. It must be of convenient cross section to seal on to the glass tubing of the letter proper. It must not be too thick, or else the high heat to which it is subjected during bombardment

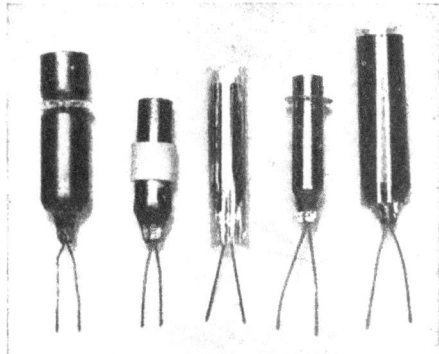

Fig. 13.—Electrode insulation. Left to right: glass beads, ceramic collar, corrugated mica, mica ring, and mica cylinder.

will crack it; and it must not be too thin, or it will be mechanically weak.

The Patented Electrode Now Fully Available. - In 1915, Claude patented his discovery that at least 1.5 square decimeters (about 23 square inches) of untreated surface area are required for every ampere of current carried. For a 30-milliampere sign, therefore, the required area per electrode should exceed approximately 5/8 square inch for reasonably long life. This patent restrained all except Claude himself and his licensees from using electrodes of this or greater area. Actually, commercial practice requires 10 to 12 square decimeters per ampere. So-called "non-infringing" electrodes were quickly developed by Claude's competitors, using smaller surface areas but making up for the difference by the use of special chemical coatings or special designs.

When the patent expired, on January 19, 1932, Claude's discovery of limited current per unit area of electrode became public property. The fact that electrodes using this principle are the best has been proved by the fact that "noninf ringing" electrodes are now rarely sold, although the discoveries made by the competing group in its effort to defeat the effect of the patent are still used and are of great value.

How Electrodes Are Made. - Most electrode shells are made by one of two methods: die stamping or spot welding. Nearly all copper and iron shells that are die-stamped have one end of the stamped cylinder tapered to a small hole into which the lead-in wire is inserted. The other end of the shell is left open to its full diameter of from % to % inch. Some iron and most of the nickel shells are rolled from metal ribbon into cylinders, and the end of the cylinder attached to the lead-in wire by spot welding. To facilitate this welding, a tab is sometimes used on the cylinder; otherwise, the lead-in wire is welded directly to the cylinder near the edge of the metal.

The wall thickness of the shell must be sufficient to withstand heating effects. Usually the metal varies in thickness from 0.007 to 0.03 inch, depending on the type of shell used. The metal must be uniform in thickness; otherwise, when being bombarded it will heat locally in the thin spots to a high temperature, which may bring the near-by glass to the melting point, with resulting puncture.

The so-called *pertruded electrode* is made of metal specially stamped to give it a perforated surface. The practice adds greatly to the mechanical strength and heat resistance of the design.

Processing the Shell. - It has been found that if the shells are specially processed during manufacture, their usefulness and uniformity will be definitely improved. Preheating in a hydrogen furnace serves to eliminate many of the impurities and gases which have been absorbed or occluded in the metal. This process does not make bombarding on the pumps unnecessary, however, since the shell quickly picks up impurities after the hydrogen firing. One of the most widely used treatments used today is that of oxidizing the metal shell. In copper shells, the oxide coat will reduce the tendency to sputter, and in iron much less tendency either to sputter or to rust is the result.

Borating. - A permanent deep-red coat of copper oxide can be formed on the shell by the process known as borating. The copper electrode is dipped in molten borax, which quickly coats it with the proper thickness of oxide. Other chemical treatments are used to avoid overheating, but they are patented processes or trade secrets not generally available to the public.

Processing to Reduce Cathode Fall. - When the electrode must be smaller than the regulation size dictated by Claude's law, an abnormally high resistance develops around the electrode, and the positive-ion bombardment gives rise to excessive sputtering. To meet these conditions, special shells have been developed by coating the metal base with some electron-emitting substance such as barium oxide or caesium oxide. These substances, used in radio vacuum tubes for the same purpose, have the property of giving off large quantities of free electrons. These negative electrons, liberated in the space surrounding the electrode, have the effect of acting as a defense against the positive-ion bombardment, by counteracting the high voltage drop around the electrode. The methods of applying the barium or caesium differ according to the manufacture, but a common method is that of coating the electrode with a barium-strontium salt held together with a butyl acetate binder, which can be easily applied. Under the influence of heat during the bombarding on the pumps the salt changes to oxide.

The Lead-in Wires and the Seal. - The lead-in wires are attached to the shell either by being pinched inside the closed end of the shell or by being welded directly to the metal. In a pinched external-seal electrode, the metal shell with the lead-in wire attached is sealed directly in the electrode glass casing. In a flare- or external-ring seal electrode, the lead-in wires are inserted in a stem machine and sealed into a glass flare, making a complete stem. The metal shell is fastened to wires on the stem, and the whole unit sealed into the electrode glass casing.

One type of external seal, known as the *metal-to-glass joint,* shows great promise. This seal consists of a cylinder of glass just large enough to fit into a cylindrical metal cap. The metal edge is beveled down to 0.005 inch to meet the glass, and the metal has a special coat which will adhere to the melted glass. By heating the joint, the glass will melt and adhere to the metal cylinder. No lead-in wires are required. This seal has been used to some extent in the neon field; the drawback to more general use is the difficulty in making every metal-to-glass seal perfect.

For use in special applications, seals made of pyrex glass have been developed. This glass must be used with a tungsten lead-in wire, which is very brittle and hard to handle. Pyrex glass, however, must be used for the electrodes if pyrex-glass letters are used, since pyrex will not seal with common lead glass. Because of the stiff, inflexible nature of the tungsten wire, copper pigtails are often fastened to the lead wires to avoid breakage in connecting and handling the electrodes. Slow leaks sometimes develop in tungsten-pyrex seals because of the presence of microscopic cracks in the tungsten wire.

One other general design of electrode was developed in the attempt to circumvent Claude's patent. This electrode has its shell covered with an insulating material, with the exception of a small opening where the discharge enters. It was believed that while sputtering goes on constantly in such a shell, the disintegration is confined to the inside of the shell; and since the sputtered metal cannot escape, a continuous agitation takes place. Any confinement of gas by the sputtering action is thus purely temporary.

Insulation Materials. - The necessity for some sort of inert, heat-resistant, and electrical insulating material to support the electrode shell away from the glass tube has already been mentioned. Many materials have been proposed for this purpose, and at least four different materials are used. These materials are mica, glass, porcelain, or other ceramic material and types of insulating laboratory cement. These materials are used in two different ways. The first makes use of a supporting member, or collar, which centers the shell in the glass tube but leaves most of the area of the shell free and open to the natural space insulation. This type of support is illustrated in Fig. 13. The second method makes use of a complete insulating shield between the shell and the glass, the shield extending over the entire area of the shell.

The first method, that of the support member, was introduced in Claude's original work; he used rings of glass beads fastened around the outside of the electrode shell. More modern forms include mica disks, ceramic and porcelain collars, etc.

The second method, that of complete shielding, is sometimes accomplished by the use of a complete coating of insulating material which covers the outside of the shell. More common is the use of the mica shield, which can be of several forms. The simplest is a plain cylinder rolled from a sheet of mica and inserted inside the glass jacket. Corrugated mica is often used to form this cylinder, since the corrugations hold the metal farther away from the glass and at the same time provide complete insulation. Sometimes a glass cylinder is used to inclose the shell. Often combinations of the insulation coat and a mica or glass cylinder are used together. A great deal of ingenuity has been used in designing electrode shell insulators, but all designs perform the same function, that is, keeping the hot and electrically charged electrode away from the glass jacket which surrounds it.

The materials used for this purpose must, of course, contain no impurities which may be liberated inside the tube after it has been pumped. Mica tends to absorb and hold moisture, as do some of the ceramics. For this reason, furnace heating is desirable before these parts are attached to the electrode. Thorough bombarding of the electrodes while the tube is being pumped is important, of course.

Hot Cathodes. - For low-voltage luminous tubes, the necessary electron supply is provided by what is known as a hot cathode. This electrode contains a wire or metal ribbon and is usually coated

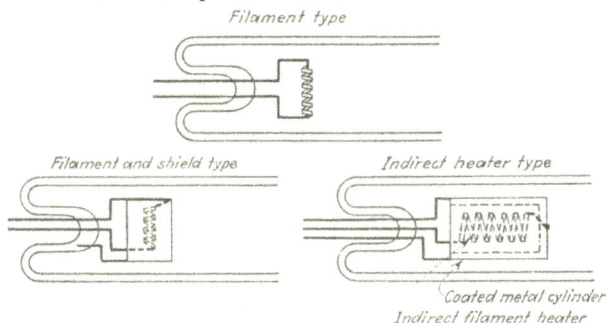

Fig. 14.—Different forms of heater-type (hot-cathode) electrodes.

39

with barium or caesium oxide. The ends of the ribbon are connected to two lead-in wires, which connect the ribbon or filament through the seal to a battery or low-voltage heating transformer. When the filament is heated to dull-red heat, the barium-caesium coating liberates quantities of free electrons which neutralize the cathode fall around the filament. This fall, which causes most of the voltage drop through the tube, can be reduced to a very low value by this means, and, as a result, the tube can be lighted at very low voltages (as low as 110 volts). Such tubes work at much higher amperages than are used in the high-voltage types.

Efforts to perfect such hot cathodes have recently met with considerable success. Although they have not been adopted today to any great extent by the sign-making trade, because the technique for using them has not been sufficiently simplified, it is expected that the hot cathode will assume considerable importance in the industry. Several forms of hot cathodes are shown in Fig. 14.

Glass Tubing

The glass used in neon tubing must have many well-defined properties. It must be strong, and it must be permanent, even when exposed to the weather and to great heat. It must melt at a convenient temperature, so that it may be worked easily in the gas fires ordinarily available. It must not contain impurities, and it must be uniform, both in diameter and in thickness of wall.

Two general kinds of glass are used in the neon-tube trade: lead glass and pyrex. Lead glass, being the cheaper and more easy to work, is by far the commoner of the two. Almost all the glass used for luminous tubes in this country is manufactured by one organization. [4]

Lead Glass. - Lead glass contains a considerable amount of lead oxide, from which it derives its name. The glass is made from silicon oxide (sand) and lead oxide melted together with potash or other substances and carefully cooled. Although glass tubing has been drawn by hand for many years and is still hand drawn for many purposes, most of the lead-glass tubing used by the neon trade is drawn in automatic machines. These machines whirl the molten glass and draw it into tubes which, after cooling, are mechanically sorted for size. The tubing is held within narrow tolerances as to diameter and thickness of wall, the maximum diameter variation being half a millimeter; and the maximum variation in wall thickness, 0.015 inch. The wall thickness in lead glass varies from 0.045 to 0.060 inch. The designation of this glass is known as G 1. It costs approximately twenty-three cents per pound.

Lead-glass tubing can be bent and shaped with ease over an ordinary illuminating gas flame, if the flame is fed with air under pressure. [5] The glass is easy to anneal (annealing is the process of slow cooling which eliminates internal strains). If too much gas is applied to lead glass, the lead becomes metallic and tends to discolor the glass, but proper reheating can usually remedy this defect. When properly handled, the glass gives very little trouble on the fires.

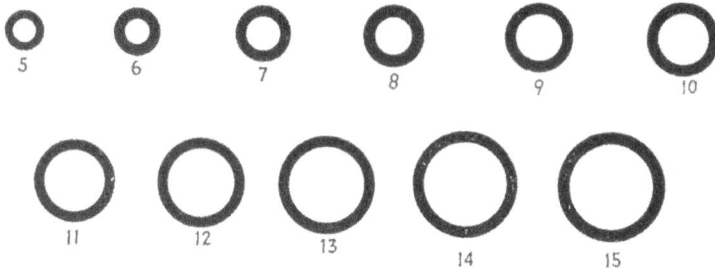

Fig. 15.—Actual size cross sections of standard neon glass tubing from 5 to 15 millimeters outside diameter, with ¼ millimeter tolerance for variations.

In common with most types of glass, lead glass absorbs many impurities, which must be removed before the tube will operate satisfactorily. These impurities consist mostly of moisture and carbon dioxide, which cling to the surface of the glass in a very thin but fast-clinging layer. Only by heating the glass in a vacuum to a temperature of at least 400 degrees centigrade will this layer be released, and the resulting free gas be pumped out. The bombarding process in tube manufacture usually accomplishes this, but if the glass is not brought to this temperature (400 degrees centigrade), trouble from moisture and carbon dioxide will almost inevitably follow. There are other impurities within the glass itself, which under ordinary conditions of operation do not interfere with the discharge. Inferior glass, however, may release these inner impurities, and for this reason only the best grade of glass should be used.

Pyrex Glass. - Pyrex, a special glass manufactured by the Corning Glass Works, has particular properties which make it desirable for neon-tube use, but it must be worked with an oxygen-fed flame, and it is relatively expensive. The glass is composed of silicon oxide, boric oxide, sodium oxide, potassium oxide, calcium oxide, aluminum, and iron oxides, with a small percentage of arsenic oxide, in the proper proportions. The glass closely resembles lead glass in outward appearance, although a pyrex tube can usually be identified by looking at the end of the tube, which shows a definite brown tinge. In its physical properties, pyrex is very different from lead glass. It is highly resistant to sudden temperature changes (it will rarely crack because of sudden cooling), and it is mechanically stronger than other glasses. It will not discolor when overheated, as lead glass will.

Pyrex tubing is drawn entirely by hand, so it is not so uniform as machine-made lead-glass tubing in diameter and wall thickness. The diameter variation is 1 millimeter; and the wall-thickness variation 0.030 inch, running from 0.040 to 0.070 inch.

Much hotter flames are required to work pyrex glass than lead glass, but if the proper equipment is available, pyrex gives more generally satisfactory results, since it can be worked easily, will not crack even though mishandled on the flame, and is unusually strong. Pyrex glass sells for about ninety cents

41

per pound. The moisture and carbon dioxide layer which cling to the surface of the glass must be removed at a much higher temperature than is necessary for lead glass. The glass must reach 500 degrees centigrade during bombardment before these impurities can be released.

Because of the differences in the thermal expansion of lead glass and pyrex, the two glasses cannot be fused together. Since a neon-tube manufacturer would often like to combine these two glasses, this fact is often a source of great inconvenience and difficulty. In order to join the two kinds of glass, either a special graded-glass seal or a ground-glass joint must be used. The graded seal contains seven different grades of glass within a length of ½ inch, which graduate in seven steps from the pyrex type to the lead-glass type. The ground-glass joint connects a piece of pyrex to a piece of lead glass by means of a ground-glass seal. The seal is made vacuum-tight by the use of a special grease.

TABLE IV.—TYPES OF SOFT GLASS

Type	Code
Clear glass	G 1
Violet absorbing soft-yellow glass (noviol)	353
Soft canary (uranium)	333 or 372
Soft red	240
Soft blue	532
Soft opal	633
Amber, medium soft	335
Soft signal green	440
Soft black	504
Soft dark purple	552

Numbers refer to trade designations of the Corning Glass Works, Corning, New York.

Colored Glass. - Since the colors obtainable from gas discharges are practically limited to red (neon), pinkishwhite (helium), and blue (mercury-argon-neon), other colors can be obtained only by using colored glass, which will combine with the discharge colors to give a wide variety of shades. Many varieties of colored glass are available to the trade.

Colored glass is usually of the lead-glass variety, with metallic oxides added to give it the desired color. Since these glasses are of lead-glass composition, they may be worked with a simple air flame and may be spliced readily with clear lead glass. But the metallic oxides which give the color make the glass brittle, and it is hard to handle. Much of this colored glass is so brittle that only expert bending, splicing, and annealing will make a successful tube. Opal white is a particularly difficult glass in this respect. All colored glass is hand drawn and for that reason is subject to the same variations as pyrex, and it is as costly.

The most commonly used colored glass is the yellow glass known to the trade as "noviol," which is used with mercury-argon filling to produce the popular green light. Uranium-oxide glass is used to a somewhat smaller extent for the same purpose. No viol glass is also used with helium gas to produce the yellow effect popularly known as "gold" tubing. Noviol is technically known as violet-absorbing soft-yellow glass, and its trade number (in the

Corning designation) is 353. Uranium glass, a "soft canary," is actually light green in color and is used to produce green tubing from the argon-mercury gas.

Other Glasses. - There are many other glasses available (see Table IV). Lime or soda glass, a very common form, is not used because of its brittleness, while the other glasses having more suitable properties cannot be considered because of the expense of manufacturing them.

Glass for tubes is usually sold in 46-inch lengths, by the pound, either directly by the manufacturer or through a supply house. It should be stored so that dust cannot accumulate inside it, and it must not be allowed to stand too long before being used.

One other form of glass is occasionally used. It is a black opaque glass, of the lead family, which is used for "crossovers," that is, the sections of the tubes which join two or more letters together. This gives a very neatappearing and permanent job, but it is much more complicated than the use of paints or tapes which accomplish the same effect.

Crossover Paints. - When two or three letters are joined together in a single tube, the connecting portions which do not form a part of the letters must be blackened out, or they will destroy the clear-cut appearance of the sign. These crossovers are most often blackened out by the use of special nonmetallic and long-wearing paint, particularly suited for coating glass. Almost any paint may be used for this purpose, of course, even the metallic ones, but if metallic paint is used near the electrodes, it will conduct the high voltage out on the sign, often with disastrous effects; and on skeleton signs, metallic paints will cause static discharges between glass crossovers. A simple, but not so neat, method consists of winding a black tape around the crossover sections.

[1] It should be remarked that there is no difference in penetrating power between red light produced by neon tubes and red light produced by any other source. F. C. Breckenridge and J. E. Nolan, writing in the Bureau of Standards Journal of Research (Vol. 3, p. 24), conclude on the basis of many experiments "that there is no difference between the visibility of light from a neon lamp and the light of the same color and horizontal candlepower distribution from an incandescent lamp." Breckenridge, writing in the Transactions of the 'Illuminating Engineering Society (Vol. 27, p. 232). states that the transmission of red light was found to be superior to the transmission of blue light in all experiments with natural fogs and on nights of low visibility. But it is not clear that red light is more penetrating than white light. (See footnote concerning the use of neon for aviation work.)

[2] Gas-discharge tubes are used occasionally for illumination purposes. These units usually combine neon and mercury-blue tubing or a combination of mercury and incandescent lights. Hot cathodes are used in tubes of this type, in place of the cold electrodes used in neon tubing, and the operating currents are much higher.

[3] Some historians claim that Ramsay named the gas from neo, meaning "related to," and the letter N, the symbol

for nitrogen. If so, neon means "related to nitrogen," a fitting name, since it displays many of the physical characteristics of that gas.

[4] Corning Glass Works, Corning, New York.

[5] A glass rod (G 1 glass) 1 millimeter in diameter will start to elongate under its own weight at 620 degrees centigrade.

Chapter Four - Electrical Equipment

Electrical Accessories for Tube Signs. - The luminous tube requires an electrical power supply, and since the supply must be of high voltage, particular care must be taken with the insulation and mounting of the various electrical accessories. Under the general heading of electrical equipment, we may list the following:

1. Insulators and mounting parts, including elevation posts for mounting the tube proper, housings or rings for shielding the electrodes, cable-support plugs for mounting the high-voltage cable, etc.

2. The high-voltage cable, which connects the transformer terminals to the lead-in wires of the tube or to the housing to which the electrode fits.

3. The high-voltage transformer, which "steps up" the low voltage from the 110-volt lighting circuit to the high voltage required for the sign.

4. The low-voltage wiring, including BX cable, low voltage switches, connections to junction boxes, etc.

5. Special switching and animating devices, such as flashers, spellers, time clocks, effect switches, photoelectric relays, of both the high and low-voltage variety.

6. Rotary converters for obtaining a supply of alternating-current where only direct current is available from the power lines.

7. Power-factor correctors, for improving the power factor of the high-voltage transformer. (In most cases, the power-factor corrector is supplied, complete with the transformer.)

8. Radio-interference eliminators, for reducing the radio noise caused by sign installations.

If all of these types of equipment are understood by the sign engineer, he will have little trouble in handling the electrical problems which arise in sign making and installation.

One important piece of equipment used in the manufacture of signs will not be treated in this chapter. It is the bombardment transformer and its regulator, devices not used as accessories to the completed sign but of fundamental importance in the making of the tubes. This equipment is described thoroughly in the chapter on bombardment.

Insulators and Mounting Parts. - The elevation post, or "neon-tube support," is an insulator used in mounting the tube on the sign box. At the bottom of the post is a metal bracket, which contains a hole for screwing to the

44

body of the sign. The total height of the post varies according to its use, from about 15/8 to 2½ inches. At the top of the insulator, provision is made for gripping the glass tube securely. In the snap-in type, a wire clasp is provided into which the glass tube fits. In the standard type (the more common of the two), the glass or porcelain insulator contains a semi-circular recess in which the tube is held by a thin wire. This wire is wrapped around the tube and passed around flanges on the post or through a small hole in the post.

In actual use, two or more posts are used to grip each separate letter or section of the tube. To keep the tube parallel to the base on which it is mounted, the posts must be of the same height. Under certain mounting conditions, it is necessary to have one post shorter than the others in order to keep the tube parallel to its background. For this reason, elevation posts are usually made adjustable in height, by means of the thread on the post, which fits into coiled-wire support, or by a solid metal piece with notches for gripping the thread. By turning the post in this screw thread, its height may be adjusted to meet the requirements. Special extension posts are used in window-border signs.

Electrode Housings. - Housings are made from dry-process or wet-process porcelain or from pyrex. The housing fits into the sign box and is mounted rigidly in place. The tube is so placed that when held by the elevation posts, the electrode jackets fit into the housing. The phosphor-bronze spring inside the housing bears against either the electrode lead-in wires or a metal cap at the tube end for making contact directly with them. The extreme outer end of the housing is fitted with a brass binding post to which is attached the high-voltage cable. The Underwriters' Laboratories have set up standard specifications (see Appendix I) for these housings, and it is always advisable to use housings which have Underwriter approval.

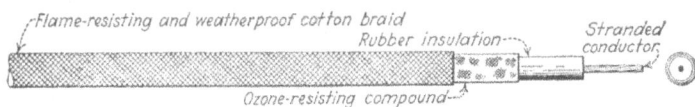

FIG. 16.—Construction of standard neon-sign cable

Electrode Rings. - In cases where a complete housing is not required, a ring can be used, which merely insulates the electrode from the metal sign. With this form of insulation, it is necessary to provide openings in the metal sign to permit access for connecting the electrodes to the high-voltage cable. Porcelain cable plugs (see Fig. 8) are used inside the sign box to support the high-voltage cable away from the metal box.

High-voltage Cable. - Attached to each electrode lead wire is a piece of high-voltage cable which connects it with the high-voltage transformer. This cable must meet very rigid specifications if it is to stand up under all weather conditions and under the heavy electrical forces that it must handle. A cutaway view of the usual cable is shown in Fig. 16. Owing to the small current carried, the Underwriters have approved the use of one size, No. 14. The wire

45

conductor which carries the current is made up of many strands of fine wire (nineteen strands are used usually), each covered with a thin coating of tin. The individual tinned wires are wrapped into one stranded wire to give needed flexibility to the cable. The wire is then coated with a thick coating of specially prepared rubber, which is then vulcanized under high-pressure steam. This rubber is treated to give it unusually high resistance, especially to high-voltage effects. Over the rubber coating is woven a cotton-braid cover. The braid is then treated with a fire-proofing solution which renders it non-inflammable, and finally many layers of flame-retarding lacquer are baked over the fabric covering. Each coat of lacquer is allowed to dry individually, so that no shrinkage will occur when the cable is complete. This grade of lacquered cable is used for skeleton window signs wherever the cable is exposed, since it presents a neat, clean appearance. Inside the sign box, however, a high finish is not necessary. For this purpose, a cable known to the trade as *mica cable* is used. In most respects, it is similar to the lacquered cable, but it has a plain braided covering. As small an amount of cable as will conveniently do the job is used, since excessive cable lengths have a harmful effect on the transformer.

Sign cable must withstand a very destructive effect known as *corona discharge.* When high voltage (15,000 volts) is applied to a bare conducting wire in air, a flamelike glow of pale violet color will surround the wire for 1/8 inch or so. This glow is called corona; it results from a direct ionization of the air. Corona has a very destructive influence on poor instating materials and will make rubber insulation brittle and useless unless the rubber is specially prepared to withstand its effects. For this reason, among others, even well-insulated ordinary electric wire is not suited to sign use. Only standard high-voltage cable especially designed for sign use should be used. Underwriters' specifications have been set up for cable, as given in Appendix I.

Cable is available for use at three different maximum voltages. The 5000-volt cable is identified by a red thread in the wire strands; 10,000-volt cable, by a yellow thread; and 15,000-volt cable, by a blue thread. The voltage rating of the cable should never be exceeded if satisfactory service is expected from it.

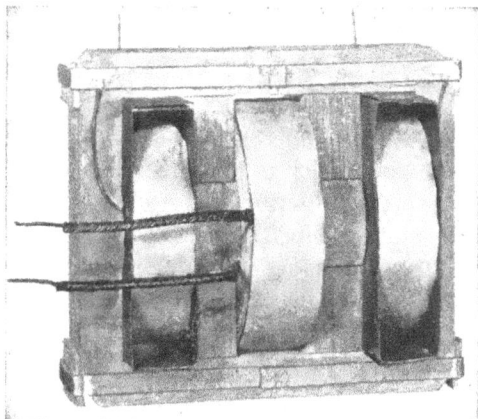

FIG. 17.—Internal view of neon transformer construction. The center coil is the primary; the two outer coils are the two halves of the secondary. The iron sections between the coils are the magnetic shunts which limit the current output.

Electrical Transformers. - The ordinary type

of transformer used in electrical practice consists of an iron *core,* or central portion, made up of thin sheets of special iron. This core usually is in form of a rectangle, with a rectangular cross section. A typical transformer core is shown in Fig. 17. On the core are wound two (sometimes more than two) windings of insulated wire, each winding consisting of many turns. The electrical action of the transformer consists in stepping up or stepping down the voltage. That is, if a voltage is applied to one winding, another voltage will appear in the other winding (although there is no direct metallic connection between the two). If the second winding has more turns than the first, the voltage across the second will be greater than that across the first winding; if it contains fewer turns than the first, the voltage will be lower. By properly proportioning the number of turns, the step-up or step-down ratio of the transformer can be made to suit almost any need.

The action of the transformer depends upon the magnetism which is set up in the core by the primary winding. The lines of magnetic force in the core link the primary and secondary together, and as a result a voltage appears in the secondary, provided, of course, that the voltage is alternating. *A transformer will not operate on direct current. If connected to direct current, it will be almost immediately burned out.*

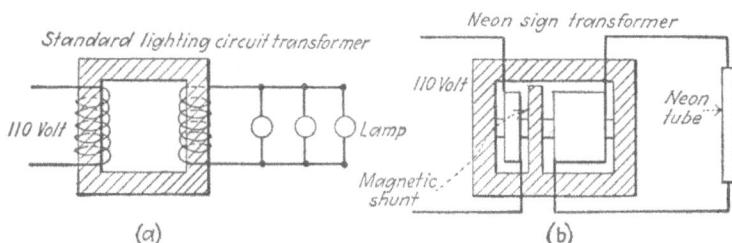

FIG. 18.—(a) Type of transformer used on standard lighting circuits. (b) Neon transformer, showing location of magnetic shunt which limits current output.

Neon Transformers. - For neon-sign work, of course, a step-up transformer is required. The primary, or low voltage winding, consists of perhaps a hundred turns of moderately heavy wire. The secondary (or high-voltage winding) consists of many thousands of turns of smaller wire. For a 15,000-volt transformer, the secondary would have approximately 140 times as many turns as the primary.

Ordinary transformers not used in neon work possess what is known as good *regulation;* that is, the voltage across the secondary is maintained almost constant regardless of the current being drawn from the secondary. This is a desirable characteristic in ordinary power work. For example, if the distribution transformer which supplies the 110-volt current to homes did not possess good regulation, the lights in the house would become dimmer as more lights were turned on.

In luminous-tube work, however, good regulation is not desirable. This arises from the fact that the higher the current through the tube the lower its resistance becomes. If a luminous tube were connected across a transformer having good regulation, the current would increase. As it increased, the resistance of the gas would decrease, and, as a result, still more current would flow. In a short time, the current would become so excessive that either the transformer would burn out or the tube would be destroyed.

What is needed is a transformer having *poor* regulation. That is, a neon-tube transformer should be built in such a way that as more current is drawn from the secondary, the secondary voltage goes down. Such a transformer, connected to a gas-discharge tube, soon reaches a condition of equilibrium, since if too great a current is drawn, the voltage goes down almost to zero, and the increased current can no longer flow.

Transformers of this type can be made easily if an extra iron path is provided in the core. This extra transformer leg is called a *magnetic shunt;* it acts as the safety valve of the transformer. As the current in the secondary increases, more and more of the magnetic lines of force are by-passed by this extra shunt, and as a result less of the lines link the secondary winding. Thus, as the current in the secondary increases, the voltage across the secondary decreases, and the required poor regulation characteristic of the transformer is obtained.

The action of the transformer can be illustrated most clearly in graphical form. Figure 19(*a*) shows the voltage appearing across the secondary of the transformer for each value of current flowing through the secondary. It will be seen that when no current flows (the so-called open-circuit condition), the full voltage of 15,000 volts appears in the secondary winding. If now more and more current is taken from the secondary, the voltage decreases. At about 25 milliamperes the working current for which the transformer is designed to operate the voltage has dropped to a much lower value. Finally, if the two terminals of the secondary winding are connected together (the so-called *short-circuit condition*), the maximum current that can flow, which in this case is 30 milliamperes, will flow. Thus the secondary of a neon-sign type of transformer can be short-circuited without harm.

The dropping voltage characteristic of the transformer depends on the inductive action of the magnetic

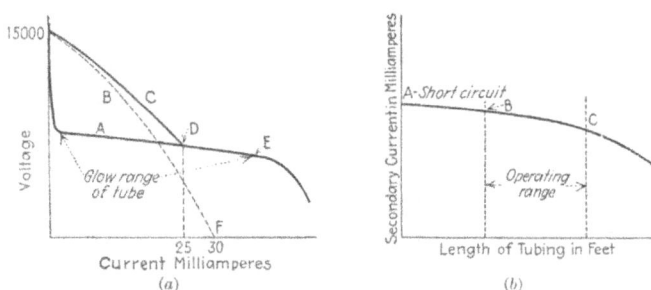

Fig. 19.—(*a*) Curves illustrating the action of a neon-tube transformer. *A* is the tube-characteristic curve showing high initial break-down voltage. *B*, transformer-regulation curve (if load were non-capacitive). *C*, actual transformer-regulation curve, shifted by capacitance of tubing. *D*, stable operating point, due to current being limited by transformer regulation. Operating current at this point is approximately 17 per cent less than the short-circuit current shown at point *F*. *F*, short-circuit current from high-voltage winding. (*b*) The current regulation curve of a typical neon transformer. Between points *B* and *C* is the safe operating range.

shunt. When the sign and its leads are connected to the transformer, the elec-

48

trostatic capacity which these wires and the tube have will neutralize some of the inductive action of the transformer. As a result, the voltage will drop off less rapidly as the current increases, as shown by line C in the diagram (Fig. 19(a)). If lead-sheathed cables are used to connect the transformer to the sign, or if other metallic coverings are used, the capacity of the external line circuit will become still greater. [1] Under this condition, the voltage of the transformer may rise to higher values than are intended, and the transformer may burn out. It is recommended by the transformer manufacturers, therefore, that no metal-covered wire be used in the high-voltage circuit and that the length of tube used be not above the maximum recommended. Otherwise, transformer burnouts will occur, and frequent replacements will be necessary.

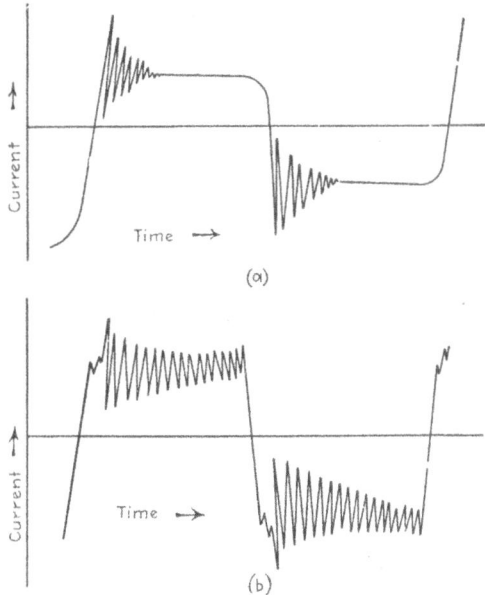

Fig. 20.—(a) Secondary current of neon transformer, properly loaded. (b) Current when transformer is overloaded. The tubing is then near the flickering point.

The leakage-reactance type of transformer, as the type of transformer described above is called, has a very poor power factor; that is, the apparent power in volt-amperes as measured by the product of the primary voltage and primary current is much greater than the actual power in watts consumed. Since meters used in electric lighting circuits measure the true rather than the apparent power, this fact is not of great importance except in special cases. In direct-current districts where rotary converters must be used, and in sections where the power company offers a premium for good power factor, it is desirable to correct the power factor of neon transformers. The method of doing this is described in a later paragraph. Most modern transformers for neon-sign use are provided with power-factor correctors mounted in the same casing with the transformer. These transformers are only slightly more expensive than the noncorrected type, and they are well worth the additional cost.

Transformer Ratings. - The power which a transformer can deliver is limited by its size. It is important to buy the size of transformer best suited to the job, as a too-large transformer is expensive in the first place and does not give good service when connected to too small a load, while the too-small transformer will not light the sign without overheating and subsequent burnout. Transformer ratings, which show the type of work for which the transformer is fitted, must be considered carefully in ordering transformers from the manufacturer or supply house.

Transformers are rated by (1) primary voltage (usually 110 volts for lighting circuits but may be 220 or 440 for special circuits); (2) primary volt-amperes (the product of the primary volts times the primary amperes when operating into the proper load); (3) primary watts (the actual power consumed when operating at the proper load); (4) secondary volts (on open circuit); (5) secondary current in milliamperes for the correct operating load (called operating current); and (6) secondary current in milliamperes for short circuit (called short-circuit current). Transformers are often specified briefly by two values: their secondary voltage and short-circuit current, since these values will usually fix its size; but all of the above factors must be considered in designing the sign installation.

The ranges provided in luminous-tube transformers are:

Primary volts: 110 to 440 volts.

Primary amperes: up to 8 amperes.

Primary volt-amperes: up to 850 volt-amperes.

Primary watts: up to 410 watts.

Secondary voltage: from 2,000 to 15,000 volts.

Operating current: from 16 to 50 milliamperes.

Short-circuit current: from 18 to 60 milliamperes. For each size of transformer, there is a definite limit to the number of feet of tubing which can be handled. This limit depends on the kind of gas used and the diameter of the tubing. Table VI gives the maximum footage allowable for each type of transformer. The footages given are for neon, mercury blue (or green) tubes using argon carrier and for

FIG. 21.—Internal construction of low-voltage type (2000 to 5000 volts) neon transformer. Only one magnetic shunt is used, whereas two are usually used in the higher-voltage types (see Fig. 17).

helium tubes. The methods of matching the transformer to the tubing and of testing for proper loading are treated in detail on pages 215 to 218.

It is always desirable to choose a transformer to fit the amount of footage and to use nearly the maximum amount of footage allowable for that transformer. In this way, it is fairly safe to assume that the transformer is neither overloaded nor underloaded. Actually, of course, the number of feet which can be operated from a given transformer is adjustable, since the correct current will flow almost regardless of the length of the tube, so long as the upper and lower limits are not exceeded. Figure 19(b) shows the secondary current for different lengths of tubing within the operating range.

Physical Construction of Tube Transformers. - The physical construction of the luminous-tube transformer depends upon its intended use and also to a certain extent upon the manufacturer's design. The windings must be thoroughly insulated from one another, and for this reason layers of heavy paper are used between each layer of each winding.

For low-voltage (2,000 to 5,000 volts) types, one magnetic shunt is used (see Fig. 21), and the secondary winding is completely insulated. For the higher-voltage (up to 15,000 volts) types, two shunts are used in a balanced construction, and the middle point of the secondary

FIG. 22.—Typical neon transformer in case, showing insulating bushings for the secondary terminals.

winding is connected, or "grounded," to the case of the transformer. In this way, the maximum voltage of either high-voltage terminal is never higher than half the terminal voltage above ground, which reduces the strain on the insulation and balances the high-voltage circuit.

The coils, wound on the core, are fitted with leads, and the whole transformer inserted in a metal case. The case is then filled with a special insulating compound which completely incloses the transformer. This filling provides in-

FIG. 23.—Core-and-coil type transformer, showing open construction, suitable only for indoor and well-protected installations.

sulation, it keeps moisture away from the windings, and it also assists in keeping the transformer cool during operation. The metal case is fitted with low-voltage terminals, which connect to the low-voltage primary winding; with high-voltage bushings to which are attached the high-voltage winding. In most cases, these bushings provide a complete covering of porcelain, so that no part of the metallic high-voltage circuit is exposed.

Depending upon the type of service for which it is intended, the exact form of the transformers differs. The standard type is used for mounting inside the sign housing. A "thin" type is used for small space requirements. A weather-

51

proof type can be hung outdoors, directly exposed to all weather conditions. A suspension type is supplied with high voltage cable, low-voltage wires, and a pull-chain switch for use in a store window or other display. The electrode receptacle type has special housings, not unlike the porcelain electrode housings already described, into which the ends of the tube may be fitted. In this type, the transformer case takes the place of the metal sign box and is often colored to improve its appearance.

The *core-and-coil* type of transformer is not mounted in a case and for this reason is considerably cheaper. Such transformers are used only in indoor installations where they can be thoroughly protected both from the public and from the weather. They are provided in voltages of from 2,000 to 7,500 volts but not higher. This type of transformer has not secured approval of the

FIG. 24.—A 15,000-volt transformer of the high-power-factor type. The case contains a capacitor for improving the power factor.

Underwriters' Laboratories except when used in a completed sign unit, inside the sign housing, and protected in such a way that it cannot be tampered with.

Power -factor Correction. - The low power factor of the leakage-reactance type of transformer has already been mentioned. When correction of this low power factor is required, there are several methods of accomplishing it. The correction is brought about by the use of a *condenser* or *capacitor* (an electrical device capable of neutralizing low power factor due to inductive loads). The condenser may be mounted with the tube transformer, in the so-called high-power-factor type of transformer, or it may be connected as a separate unit. The separate unit contains two terminals, which are connected across the low-voltage side of the transformer. Power-factor correction on the high voltage side of the transformer is not practicable.

The reason for power-factor correction lies in the fact that a low-power-factor transformer takes more primary current for a given power output than does a high-power factor type. The line losses, which depend on the square of the primary current, are thus much higher. Power companies, who bear the cost of the line losses, often will offer a special rate if the power factor is kept high. In large high-power installations, this reduced rate will often amount to a considerable sum, enough to justify the expense of power-factor correction. In such cases, the local power company should be consulted, and their recommendations considered before power-factor correction is undertaken. The case of rotary-converter operation, where power-factor correction is also desirable, is treated in the next section.

Rotary Converters for Direct-current Operation. - Alternating current, required for every luminous-tube sign, is by no means universally available. In many sections, particularly in the built-up business districts of the cities, direct current is provided by the power companies. In such cases, it becomes necessary to employ some sort of machine for converting the direct current into alternating current. The commonest machine for doing this is the *rotary converter,* or *dynamotor*, so called because it is a rotating machine (closely resembling an electric motor in outward appearance) which converts direct into alternating current. The rotary converter has two sets of terminals; the input terminals are connected to the direct current supply, while the output terminals, which provide the alternating current, are connected to the tube transformer. The converter rotates as a motor, and its speed of operation depends upon the direct-current voltage which is supplied. If this voltage is lower than rated, the speed of the converter will be below normal. If this is the case, the frequency of the alternating current, which depends upon the speed of rotation, will be lower than normal. If a transformer is operated on alternating current of lower than its rated frequency, the transformer will take excessive current and may burn out; in any event, the operation will be uneconomical. Hence it is important to see that the direct-current supply is up to normal and also to see that the converter is not overloaded. Measurement of the input direct-current voltage with a voltmeter and of the speed with a tachometer are both recommended.

The converter must, of course, be large enough to supply the load that it must carry. Converters are usually rated in volt-amperes, the product of the output alternating current volts times the output alternating-current amperes. The sum of the volt-amperes of all the transformers to be operated from the converter must not exceed the rating of the converter; otherwise it will be overloaded and will not come up to proper speed. Most converters are built so that they require practically no attention except occasional oiling of the bearings, and they can be made remarkably noiseless in operation. The larger sizes must be started with a step-by-step starter, although the smaller sizes (under 750 volt-amperes) may be started simply by closing the direct-current switch.

If the transformer which loads a rotary converter is of the low power-factor type, serious arcing at the brushes and commutator of the converter will occur. To prevent this from happening, two alternatives can be used. The first is the use of a special converter having separate direct current and alternating-current windings (an expensive type of machine), or else the power factor of the transformer load can be corrected by the use of capacitors or capacitor transformers. The latter method is usually to be preferred, because it is the most efficient way of doing things and not particularly expensive.

There are two other methods of converting direct current to alternating current which are occasionally used for neon-tube work. They are the vibrating-reed interrupter and the vacuum-tube inverter. These units can be bought complete and ready to install according to the instructions given by the manufacturer, and since they are so little used, no attempt will be made to go into their methods of operation here.

Low-voltage Wiring. - The wiring between the low-voltage side of the transformer and the power circuit must usually be installed by the sign craftsman, and regulation electricians' practice must be followed. The types of wiring and switching for this low-voltage circuit have been standardized by the electrical industry. The chapter on electrical practice describes the materials and the methods used.

Sequence Effects, Spellers, Flashers. - A sign operated continually, that is, "on" during the entire period of operation, consumes a maximum amount of power, but because of its static and monotonous appearance it may not attract the maximum amount of attention. Some advertisers prefer to have their signs turned on and off several times per minute. Less power is used, and the sign attracts more attention. For flashing signs, a simple automatic switch, or flasher, may be used; this device is made for installation on either the low or the high-voltage side of the transformer. For spectacular installations, of the so-called animated variety, separate parts of the sign may be lighted one after the other. Often the different letters in a sign are lighted one after another, in which case the so-called "speller" effect is obtained. Or in a sign having special figures, such as trade-marks or other shapes, separate parts of the figures may be lighted separately. The most effective tube signs today make use of such effects, and it is wise for even the smaller sign manufacturer to be prepared to install them for his customers.

One form of speller device is a motor-driven high-voltage switch. The switch itself consists of several contacts (one for each section of the sign to be lighted) arranged in a circle. A switch arm, turning on a shaft, makes contact with each of these points one after the other. It is connected to one terminal of the high voltage transformer, while the switch points are wired to the corresponding terminal of each section of the sign. A motor causes the switch arm to revolve so that it connects each section in order. The motor which drives the switch arm is usually a small disk motor. The switching is accomplished at high voltage; there is, therefore, a spark as each contact is made. The contacts must be made to withstand this arcing, and the whole switching mechanism must be carefully inclosed.

More complicated types of switching for large and complex installations can be provided by the companies which manufacture these devices. In one type of device, the primary of the transformer is disconnected as each contact in the high-voltage circuit is made; and since the high voltage is thus momentarily off during the contact period, no arcing results. Many variations of these devices are available to suit the particular requirements of each installation. In principle, they are much the same, however, and it is simply necessary to make the proper connections to secure the desired animated effect. The several companies which specialize in this type of switching device have complete bulletins on the subject which should be consulted before designing an animated sign.

Radio Interference. - Almost every type of electrical apparatus is capable of producing interference with radio programs, and the neon sign is no exception. If the length of tube operated from a transformer is too long, the sign will begin to flicker, or "bead." This produces a flickering voltage of some 10,000 or 20,000 cycles per second frequency, which, because of its high frequency, will be radiated out into the area surrounding the sign. The near-by radio sets will pick up this interference unless special precautions are taken.

Another source of interference is the switching devices used with neon signs. The arcing in either low- or high- voltage switches can cause high-frequency radiation which, if received by a radio set, will seriously interfere with the enjoyment of the programs. Arcing between connections, particularly in electrode housings, or between cable lengths is another cause of interference.

A set of rules to observe for the prevention of this sort of trouble is given in Chap. Eleven. Sometimes special equipment must be installed to correct the difficulty. This usually takes the form of a condenser made especially for interference prevention and connected across the arcing contacts. This may be particularly necessary in rotary converter installations, where the arcing at the commutator may cause serious trouble.

The radio-frequency voltage which results from overloaded transformers is a serious problem which cannot be corrected except by the use of a larger transformer or smaller sign length. It is thus doubly advisable not to exceed the manufacturers' rating on a transformer, even though it may appear that money may be saved in this manner.

[1] Many neon manufacturers have claimed that it is injurious to operate a neon transformer for any length of time on open circuit, that is, without any load connected to its secondary terminals. It is true that the voltage under these conditions is higher than under operating conditions and that the insulation is thus subjected to a greater strain, but all transformers made by reputable manufacturers have a sufficiently large safety factor to take care of this condition.

Chapter Five - The Types of Signs

Sign Backgrounds. Besides the luminous tube and the electrical equipment necessary to light it, there is another fundamental part of the neon sign which has received only passing attention thus far. It is the mounting of the tube on a sign background, so that the tube will be shown off to best advantage. A good part of the sign-making business is taken up with designing, building, and erecting these sign backgrounds, and for this reason we should now examine the various forms that they may take. The types of signs are divided into two main classes: the indoor and the outdoor.

Outdoor Signs. 1. The *Swing Sign, or Projection Sign.* The swing sign, shown in Fig. 25, is hung from a bracket support so that it may swing freely. Usually the sign is fitted with a tube on each face, in which case it is known as a double-face swing sign. Occasionally only one face is fitted with a tube. The dimensions of the sign vary widely according to the particular installation. The metal sign box usually contains the transformer. In very rare cases, the transformer may be mounted inside the building, but the long length of exposed high-voltage cable makes this an undesirable type of installation. If the sign is of the flashing variety, both faces may flash together or alternately, one after the other. The *projection, or cantilever,* sign is similar in construction to the swing sign, but it is mounted flush

Fig. 25.—Swing sign.

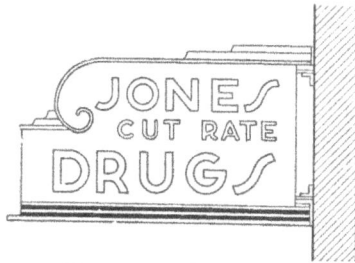

Fig. 26.—Projection sign.

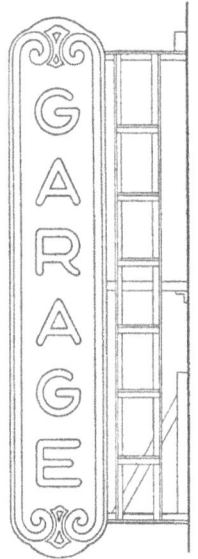

Fig. 27.—Upright sign.

on the side of the building without the use of guy bracing either above or to the sides. A typical projection sign is shown in Fig. 26.

2. *The Upright Sign.* Figure 27 shows a typical upright sign, so named for its upright position. Such signs are usually longer and narrower than the swing sign, and they are mounted by means of several brackets attached to one edge. Wire, chain, or angle-iron bracing is often used to prevent swinging. The transformers are usually mounted inside the metal box. This type of sign

is especially well adapted for large letters, visible at quite a distance, and for taking advantage of the vertical length of a building.

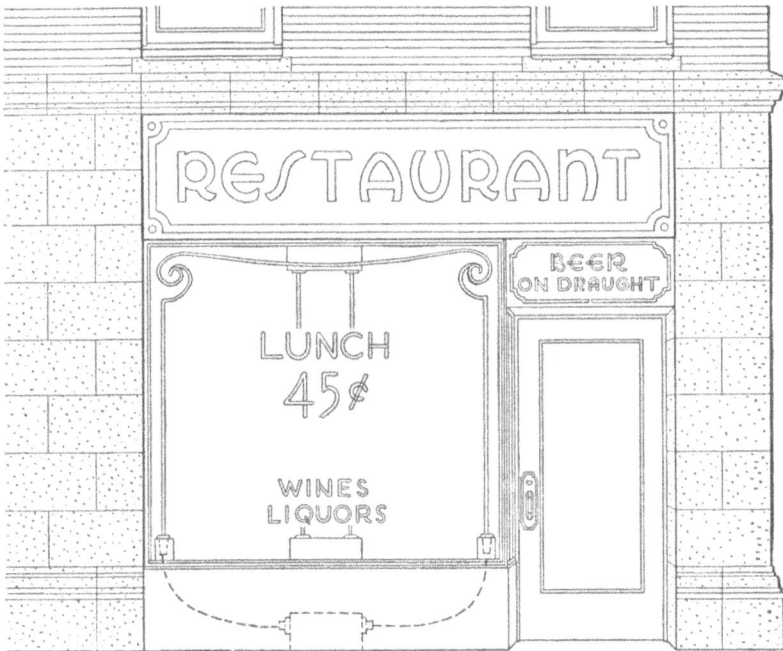

FIG. 28.—Store front, showing wall, vestibule, skeleton, and border-type signs.

3. *Wall Signs.* Wall signs, so named because they lie flat against the building wall, do not differ from the swing and upright signs in general construction but are mounted in position against the face of the building.

4. *The Outline Skeleton Sign.* Long lengths of tubing are often used to outline the edges of buildings, as shown in Fig. 29. Since the tubing is not backed up by a special background, it has the open "skeleton" effect, from which it gets its name. The transformer supply for such signs is often located inside the building, although it may

FIG. 29.—Outline skeleton sign applied as a triple-tube unit.

be mounted in a waterproof metal box outside.

5. *Roof Structures.* The roof-structure type of sign may be of two types. The open roof structure shown in Fig. 30 consists of a framework of iron angle pieces on which the individual letters are mounted, perhaps with metal reflectors backing up each letter. The closed roof structure is a complete billboard type of sign on which the letters are mounted and emphasized by reflectors. In a great many cases, this type of structure uses a combination of neon tubes and incandescent bulbs. As the name suggests, roof structures are mounted on the roofs of buildings or on other high and exposed locations.

FIG. 30.—Roof structure.

6. *The Pedestal Sign.* A box-type sign mounted on the top of a metal or masonwork pillar, as shown in Fig. 31, is known as a pedestal sign. This type is popular for use in gasoline service stations and elsewhere wherever there is no building or other support on which to mount the sign.

7. *The Vestibule Sign.* The vestibule sign is mounted outside the door of an establishment but often inside the vestibule and in the place occupied by the door transom, as shown in Fig. 28. The sign is not different from the swing or wall sign except in its location.

8. *The Marquee.* The theater-type display shown in Fig. 32 is known as a marquee. It is mounted on the special iron structure projecting over the sidewalk. It is almost always found in front of theater entrances.

In each of the above types of sign, the neon tubes may be only part of the illuminated display. Often bulb-type signs are mounted within a neon border. The removable opal-glass letter signs such as are used to announce the picture currently shown at the theater may be surrounded by a decorative neon border. Neon and mercury tubes are also being used in a solid bank to illuminate the space underneath

FIG. 31.—Pedestal sign.

the marquee, the combination of red and blue giving a brilliant display of color.

Indoor Signs. 1. *The Skeleton Sign.* The skeleton sign (Fig. 28) is a continuous tube formed into the desired reading matter. The glass tube itself is the sign. The transformer may be mounted in a metal box from which the skeleton tube hangs. Often the transformer case itself is provided with electrode housings which hold the electrodes. In this case, the skeleton sign may either be suspended from the ceiling or set into the transformer, which is placed on the window sill. These alternative types of mounting are shown in Fig. 28. In other cases, the transformer is mounted to the side, and the tube is supported on support wires, often the high-voltage cables themselves. This type of sign is popular for use in display windows, but it will serve well in any location where the tubing is not apt to come in contact with nearby objects.

FIG. 32.—The marquee sign, used by theaters and restaurants.

2. *The Box Sign.* The box sign, like the swing sign, has its tubing mounted directly on a metal sign box. The transformer is mounted inside the box, and the entire high voltage circuit is completely inclosed. The box is hung on support wires, chain, or piping of some sort or is placed on a window sill. The low-voltage connections reach the box either from the side or by way of the supports. An interior view of such a sign is given in Fig. 8.

3. *The Window Outline.* The window outline is becoming very popular largely because it has two uses: attraction to the window and illumination of the window display. The tube is of the skeleton variety and is mounted in much the same way as the outside outline skeleton sign. Often special curlicue designs are employed at the corners, and in many cases several different colors of glass are used. The transformer may be mounted under the sill of the window as shown (Fig. 28) or above the window as a hanging transformer or to the side, depending upon what is most convenient.

4. *The Display Sign.* A popular method of using neon for indoor display is the form known as the display sign. These signs are small, usually a simple border of one or more colors, perhaps with a letter or trade-mark worked into the tube. The border surrounds a message which can be inserted by the dealer, changed from day to day to meet the changing requirements of sales, menus, etc. Electric clocks are often mounted inside such neon borders.

59

Letter Types. - In box-type signs, the box itself is usually painted or treated in some way to emphasize the neon tubing so that the message of the sign will be visible when the tube is not lighted. This background lettering can be of several different types. The simplest is the *flat letter,* shown in Fig. 33 (*a*). This is simply a letter painted on the flat side of the metal box immediately behind the glass letter. The color of paint used should harmonize with the color of the tubing when lighted, so that a pleasing

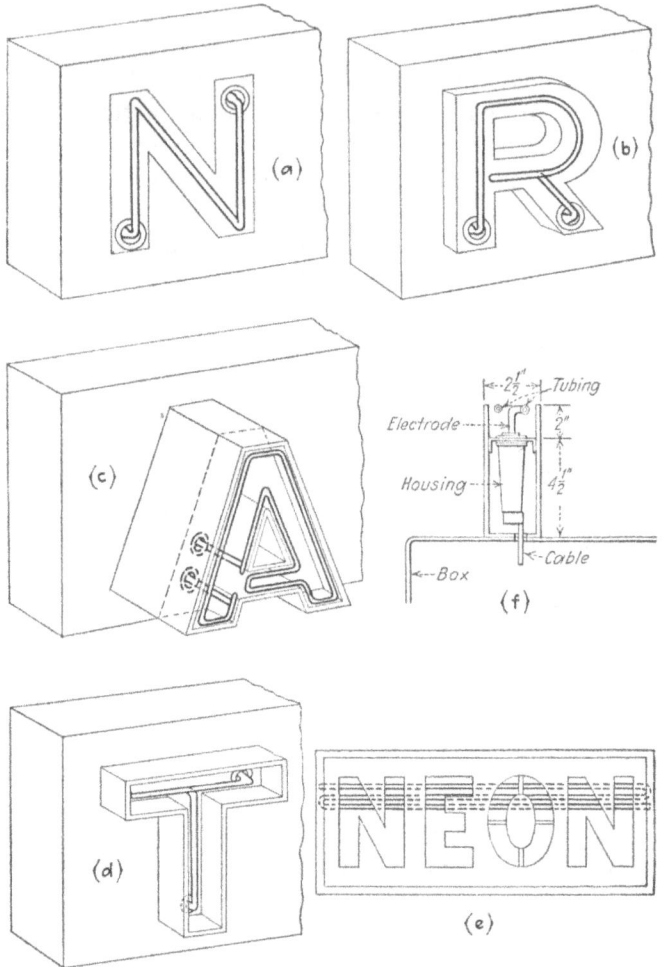

FIG. 33.—(*a*) The flat painted letter. (*b*) The raised letter. (*c*) Raised letter with shallow channel. (*d*) The channel letter. (*e*), The cut-out letter. (*f*) Section through a raised-channel letter showing electrode housing in place.

appearance is presented. This type of lettering has been found to have very little "daylight effect"; that is, it is not clearly visible in daylight, unless the tube is lighted.

The *raised letter,* shown in Fig. 33(*b*), is a simple metal letter raised above the surface of the metal box itself and having the same outline as the glass lettering. This type of letter is easily seen in daylight and gives a finished appearance. In many cases, it is made of stainless steel or stainless-steel finish. The raised letter may also be painted a different color from that of the metal box, and the same care in choosing colors should be used as in the case of the flat letter. Raised letters made of teakwood are frequently used in Europe.

The *channel letter,* also shown in the figure, is a letter outline made of metal. The tube fits into the channel, which acts as a reflector. It is of striking appearance, and it prevents the light from adjacent letters from "running together" and blurring. Such signs are very clear-cut in appearance. Channel-letter signs have a disadvantage of giving poor angle readability; that is, they are difficult to read unless viewed face-on. The raised and channel letter are often combined into the *raised-channel letter*, with a very shallow channel, which has the advantages of both types.

The *cut-out letter*, not so common as the others, consists of a sheet of metal with the letter outlines cut out from it. Neon tubes are put behind the metal and shine through the cut-out portions.

Metalwork. - The metalwork in making sign boxes involves bending, cutting, drilling, welding, riveting, soldering, and so on. No attempt will be made to de-

Fig. 34.—Letter forms.

scribe these operations in this book, since many excellent reference books on metalwork are available, and since the space required for a description of metalworking can be more profitably devoted to the particular problems of the neon industry. A few words may be said about the metals used, however. Sheet metal is universally used, varying in thickness from gauge 18 to 26 (see Underwriters' specifications). The metals are of several varieties: galvanized sheet iron, "galvaneal" (patented process of treating sheet iron to prevent rust), stainless steel, enameled iron (sheet iron with a coating of enamel baked on it, similar to the finish used on automobile bodies), black iron, and, in some rare cases, burnished copper. Black Carrara glass, ½ to 1 inch thick, is also widely used as a background material. In the construction of a metal box, it is very important to provide angle- or channel-iron braces inside the box to give the necessary rigid construction. In fairly large signs, the sheet metal itself is not rigid enough to prevent bending or twisting. In smaller signs, channel braces are constructed of the same metal used for the face of the sign and placed beside the box to give the bracing required and provide the proper supports for fastening to the face of the building.

61

Part Two - Shop Practice from Specifications to Completed Sign

Chapter Six - Designing the Sign

Sign-making Procedure. - There are many possible ways of going about the business of making a neon sign, but in general the following procedure has been found to be the most satisfactory. First, a preliminary, or "artist's," sketch of the sign is made, in more or less detail but on a reduced scale. This sketch must show the message, the types of letters, the colors, and the style of mounting to be used in the sign. It is submitted to the customer for his approval. With it is often submitted a statement showing the cost of the complete sign, the life guarantee, and an estimate of the cost of operating the sign.

From this preliminary sketch is drawn a complete mechanical drawing, or *working layout,* which gives the details of every part of the sign, including metalwork, glass bending, electrical apparatus, and connections. This layout is drawn full size on heavy drawing paper. All dimensions, locations of drill-holes, etc., are accurately located on it. The layout serves many purposes. From it are prepared the glass-bending layouts, the metalworking templates, and the connection diagram. From it is made a list of parts required, the number of feet of tubing, the type and number of transformers, the number of electrodes, and so on. The layout is an almost indispensable aid in computing costs, both of materials and of labor. And, finally, it serves as a permanent record of the sign and its construction, for use in repairing and replacement after the sign has been installed. For this last reason, the complete working layout should be carefully preserved at least until the sign is no longer in service.

After the working layout is completed, an asbestos glass-bending layout is made from it. The glass blower then bends the tubes to shape, using the glass layout as his guide. The metalwork is based on a drilling and mounting template, which is likewise prepared from the complete layout. The electrical apparatus is taken from stock as indicated by the layout list. The tubes are pumped, bombarded and filled, sealed, aged if necessary, and mounted on the sign box. The electric wiring is then completed, and the sign, properly painted, is ready for installation. Each step in this procedure is taken up in the following pages.

The Preliminary Sketch. - The preliminary, or artist's sketch, is intended to illustrate to the customer the design of the sign as it will appear when completed. It may be very elaborate, showing a detailed picture of just what the sign will look like. Such detailed sketches are very useful in capturing the customer's order by appealing to his fancy and by showing him what he is to get for his money. The sketch may be, on the other hand, a very rough outline penciled in the salesman's notebook, or it may be merely an oral description of what the customer expects the sign to look like. In every case, the preliminary plan must show the message of the sign, the type of mounting, the colors to be used, and in most cases the style of lettering desired.

The Original Layout. - It is the duty of the layout man (in small plants this man may also be the owner, salesman, glassworker, and pump man) to take the preliminary sketch and to draw up from it the complete, accurate, full-size layout from which the tinsmith, tube bender, and electrician may work. The amount of information which this layout contains varies widely in different plants, but it is recommended that it contain every possible fact about the sign. The layout man may confer with the other workers in the plant before deciding on the design, but all of the facts should appear on the layout sheet, regardless of who originated them. The layout may then be taken as the complete record for cost computation and subsequent repair work, without fear of omission of some vital feature.

With this idea in mind, it is recommended that the original layout contain the following items, either on the layout sheet itself or on specification sheets attached to it:

1. The design of the metal box, in full size, showing the design of letters, layout of metal trimming and ornaments, the metal supports for strengthening the box, angle or channel iron, and methods of fastening projecting pieces from box to building.

2. The location of all mounting holes for elevation posts, housings, cable plugs, transformers, and angle or channel iron.

3. The kind and color of the glass tubing for letters and borders.

4. The diameter of the tubing, in millimeters.

5. The length of each tubing unit, including bend backs and returns, in feet and inches.

6. The kind of gas, gas mixture, or vapor used inside each section of tube.

7. The pressure of gas inside each section of tube.

8. The location and kind of electrodes to be used in each section of tube.

9. The size and location of each transformer, flasher, speller, or animating switch.

10. The complete wiring diagram, showing high-voltage wiring, cable, jumper wires, etc., and the low-voltage wiring, switches, time clocks, etc.

Designing the Tube. - It is clear that the layout cannot be drawn up until the basic design of the sign has been decided upon. Since the entire layout of the sign depends upon the tubing, the first job to be completed is the design of the tube. By tube design is meant deciding upon the color, diameter, and length of each glass tube, and the gas and gas pressure to be used. Table V can be used for this purpose. It shows the color of glass, the gas, the gas pressure, and whether mercury is used, to produce each color, both for the standard colors commonly used and for the less common " special-effect" colors.

TABLE V.—LUMINOUS-TUBING COLOR AND PRESSURE CHART

Color of tubing	Rare gas used	Glass tubing used
"Standard":		
Red..........	Neon	Clear
Dark red.......	Neon	Soft red, Code 240
Light blue.....	Linde 50 or Airco B-10*	Clear
Dark blue......	Linde 50 or Airco B-10*	Soft blue, Code 532
Light green....	Linde 50 or Airco B-10*	Soft canary, Code 333 or 372
Medium green .	Linde 50 or Airco B-10*	Noviol soft yellow, Code 353
White.........	Helium†	Clear
Gold.........	Helium†	Noviol, Code 353 (soft yellow)
"Special effect":		
Soft red........	Neon	Soft opal, Code 633
Dark green.....	Linde 50 or Airco B-10*	Soft medium amber, Code 335
Soft white.....	Helium†	Soft opal, Code 633
Orange........	Neon	Noviol, Code 353 (soft yellow)
Red-lavender ..	Neon	Soft dark purple, Code 552

GAS PRESSURES FOR STANDARD NEON AND MERCURY TUBES
(Use with neon and Linde 50 or Airco B-10)

Diameter of tubing, mm.	Pressure for lowest voltage per foot, in mm.	Pressure allowing for absorption in use,‡ in mm.	
		Uncoated electrodes	Coated electrodes
7	17	20	18
8	16	19	17
9	14	17	15
10	12	15	14
11	10	13	12
12	9	11	11
13	8	11	10
14	7	10	9
15	6	10	8

* Used with liquid mercury in tube.
† Use at 3 mm. pressure, with special electrodes.
‡ These pressure values recommended, since they allow for one year of normal operation.
Glass code numbers are those of the Corning Glass Works, Corning, New York.

After the proper gas and glass have been chosen, the size of each section of tube must be decided upon. In a skeleton sign, the entire message is usually made up of one or two pieces of tubing. In the box sign, however, the sign is broken up into many separate tubes, perhaps one tube for every two or three letters in the sign. The separate sections are then connected in series with wire jumpers. This procedure makes the glass bending much easier to handle and greatly simplifies the mounting and repair work.

TABLE VI.—TRANSFORMER CHART

Showing maximum number of feet of tubing a given transformer will carry

Secondary volts	Secondary milli-amperes short-circuited	Volt-amperes	Operating primary watts	Primary amperes, open	Red 7	Red 9	Red 10	Red 11	Red 12	Red 15	Blue 7	Blue 9	Blue 10	Blue 11	Blue 12	Blue 15	White/gold 9	White/gold 10	White/gold 11	White/gold 12
15,000	60	875	400	8.0	...	28	32	36	43	60	...	34	38	44	54	70	..	13	18	23
15,000	30	450	210	4.0	...	27	31	34	42	58	...	32	36	42	50	68	9	12	16	22
12,000	30	350	175	3.2	...	21	24	28	33	46	...	26	29	34	39	54	7	9	12	17
12,000	25	280	145	2.5	...	18	23	25	30	40	...	22	27	31	36	49	..	7	10	15
9,000	30	280	130	2.5	...	14	17	19	22	32	...	18	20	23	28	40	..	6	9	12
9,000	18	200	80	1.8	9	12	15	17	20	28	11	15	18	20	24	34				
7,500	30	245	100	2.2	...	9	12	15	17	21	...	12	15	18	21	26				
7,500	20	150	72	1.3	7	9	12	15	17	..	8	12	15	18	21					
7,500	18	140	70	1.2	7	9	12	15	17	..	8	12	15	18	21					
6,000	30	170	88	1.5	...	...	9	11	13	17	...		11	13	15	22				
6,000	20	130	60	1.1	6	7	9	11	13	17	7	9	11	13	15	22				
5,000	30	160	70	1.4	...	...	8	9	11	15	...		9	11	13	18				
5,000	20	100	50	0.9	...	6	8	9	11	15	...		9	11	13	18				
5,000	18	95	48	0.8	5	6	8	9	11	15	6	8	9	11	13	17				
4,000	30	130	60	1.1	...	5	6	7	8	12	...		8	9	10	14				
4,000	20	75	38	0.6	...	5	6	7	8	..		6	8	9	10					
4,000	18	70	36	0.6	...	5	6	7	8	..		6	8	9	10					
3,500	18	70	35	0.6	2	3	4	5	6	..	3	4	5	6	7					
3,000	20	55	30	0.5	1.5	2	3	4	5	..	2	3	4	5	6					
2,000	20	45	22	0.4	1	1.5	2	3	4	..	1.5	2.5	3	4	5					

This chart compiled from an average taken from various charts listed by leading transformer companies in the United States. Values given are conservative, to allow for variations in glass diameters.

Metalwork Design. Once the tubing has been designed, the mechanical layout can then be prepared. For each electrode, the hole for an electrode housing or bushing must be indicated. An elevation post should be inserted for every 15 inches of tubing but never less than two to a tubing section. It is important to place an elevation post as close as possible to each electrode to relieve the shock at these vital points. At the same time, all other holes for screws, mounting brackets, etc., can be inserted in the drawing. As soon as the size of transformer has been decided upon (see paragraph below), holes for mounting it should be indicated in the proper place. In making the mechanical part of the layout, the thickness, or gauge, of the metal sheet should also be given, and attention should be paid to the mechanical strength of the entire metal box.

Electric Design. - The type and size of the transformer to be used depend on the current to be used in the tube, on the gas used, and on the length and diameter of the tubing. From the layout sheet the total length of tubing can be measured accurately with a tool known as a map-measuring wheel (see Fig. 35). If more than one color is used, the footage of each color should be measured separately. When the total footage has been computed, reference to Table VI will give the type and size of transformer required to light the sign. Figures 36, 37, and 38 are also very useful in calculating the approximate cost of operating the sign, the volt-amperes required to operate the sign, and the footage required.

Where flashers or animators are to be used, the wiring diagram supplied by the manufacturer should be studied, so that the device can be inserted in the proper place to make the wiring short and convenient. When the location has been decided, the holes for mounting the flasher should be indicated in the proper place on the layout. When all the electrical devices have been located, the wiring diagram can be drawn into the layout chart.

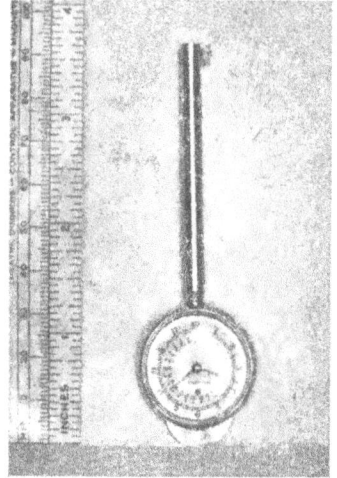

Fig. 35.—Map-measuring wheel, for determining length of tubing sections indicated on layout. As the small wheel runs over the layout, the feet and inches are indicated on the dial.

With the complete facts and figures on the tubes, the metal work, and the electrical devices on the paper layout, the layout may be considered complete, and the design of the sign is finished. The layout may then be used to draw up the specifications for glass bender, tinsmith, and electrician, as outlined in the following chapters. A complete layout chart of a simple sign is shown in Fig. 39. This diagram shows the correct layout

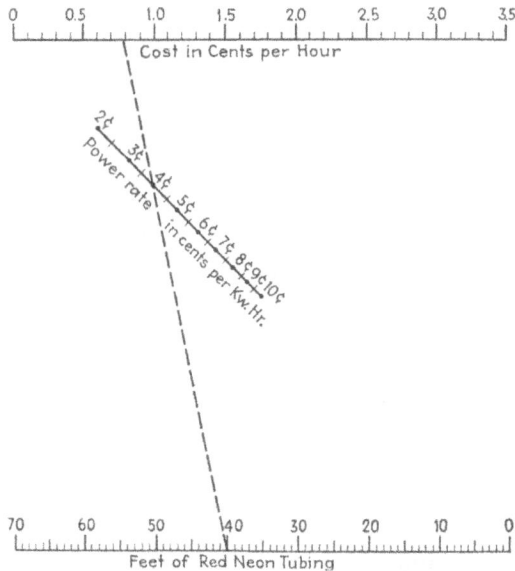

Fig. 36.—Chart for determining the approximate cost per hour. Place a rule so that it connects the number of feet of tubing and the power rate per kilowatt-hour. The point where the rule intersects the top scale will indicate the approximate cost in cents per hour. In the example given by the dotted line, 40 feet of tubing at 4 cents per kilowatt-hour is found to cost about 0.8 cent per hour.

for the complete design of the sign.

66

Preserving the Layouts. -Some provision should be made in every shop for the preservation of each layout drawing after the sign made from it has been installed, since these layouts are the only accurate record of the signs that have left the shop. They should be kept readily accessible at all times, regardless of when the sign was installed. They should be located in racks convenient for the use of employees yet out of reach of those who have no business near them. Each layout, template, and asbestos glass layout should be tagged in some logical manner, so that it is not necessary to look through the entire lot to find the one needed. A company starting in the business should not forget to provide ample space for this sort of storage, because the drawings will accumulate rapidly.

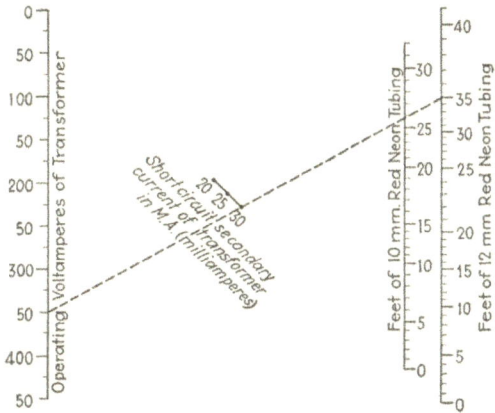

Fig. 37.—Chart for figuring the approximate volt-amperes required to operate a sign. Place a rule on the chart so that it connects the number of feet of tubing with the short-circuit secondary current of the transformer, and the rule will intersect the scale at the left at the approximate value of volt-amperes required. In the example, 35 feet of 12-millimeter tubing on a 30-milliampere transformer will consume about 350 volt-amperes.

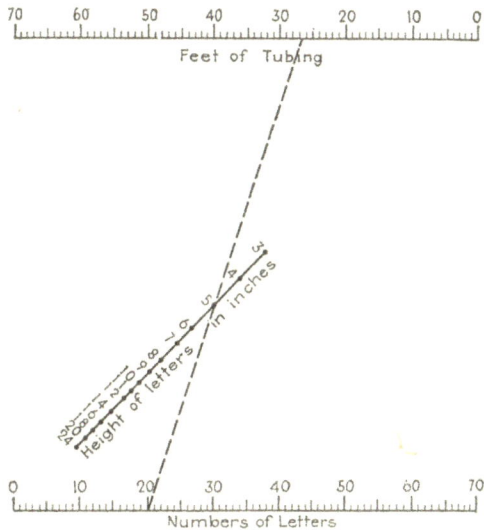

Fig. 38.—Chart for figuring approximate tube footage from the number of letters and the height of letters in inches. Place a rule on the chart so that it connects the number of letters with the height of letters in inches. The intersection of the rule with the top scale will indicate the approximate number of feet required. In the example, 20 letters, 5 inches high, require nearly 27 feet of tubing.

Cost Computation from Layout Charts. - The original layouts, metal templates, and glass-blowing layouts provide an excellent basis for computing sign costs. Practically every item which goes into the construction of a sign can be taken from a well-drawn layout. From a list of such items the cost of materials can be readily computed (see Table VII). Labor costs, both of tubing and of box manufacture, can also be estimated from layouts, since the average time re

TABLE VII.—SPECIFICATIONS FOR SIGN SHOWN IN FIG. 39

Double face swing sign reading

DRUGS [red]

SODA [blue]

with green border

flat-paint job

Glass

Reading	Size of letter, inches	Glass tubing, milli-meters	Color	Type of elec-trode	No. of elec-trodes	Gas pressure, milli-meters	Mer-cury	Linear feet, 1 side	Linear feet, 2 sides	Cost of elec-trode	Cost of glass	Cost of gas	Cost of mercury	Total cost
Drugs	10	12, clear	Red	D1	8	12	No	12.5	25.0					
Soda	10	12, clear	Blue	D2	8	10	Yes	10.0	20.0					
Border		15, Noviol	Green	D2	16	10	Yes	20.5	41.0					
							Total 1 side	43						
							Total 2 sides		86					
													Total cost	

Metal

Material	Type	Quantity	Cost
Sheet iron	No. 24 galvanized	62 sq. ft.	
Outlet box	3-in. round	1	
Flatiron	¼ by 2 in.	8 ft.	
Hanging pole	10 ft. standard	1	
Guy chain	10 gauge 2/0	30 ft.	
Expansion bolts	¼ by 2½ in.	3	
Sheet-metal screws	7 by ½ cadmium R. H.	78	
			Total

TABLE VII.—SPECIFICATIONS FOR SIGN SHOWN IN FIG. 39.—(Continued)

Electrical Material

Material	Type	Quantity	Cost
Transformer	15,000 volts, 30 milliamperes	2	
Porcelain housings	R	32	
Elevation posts	No. 1 glass	78	
Cable support	K	8	
High-voltage cable	Mica, 15,000 volts	24 feet	
No. 14 BX cable	2 conductors	35 feet	
No. 14 weatherproof	Single conductor	20 feet	
Switch (metal box)	Double pole, single throw	1	
			Total

Paint

Brush job:		Quantity	Cost
			Total

Labor Cost

Job		Hours	Cost
Glass bending			
Pumping			
Metalwork			
Electrical work on sign			
Electrical work on job			
Painting			
Assembling			
Hanging			
Sketching and layout			
		Total	

Total Costs

Material			
Labor			
Permit			
Maintenance			
Overhead			
Total			

quired for a given size box or a given footage of tubing can be found with experience. Even overhead costs can be estimated from figures taken from layouts, and in well-organized plants the entire cost structure of the business is based on layouts and layout lists. For this reason, among the others given previously, every template and layout should be complete, and it should be kept on hand while any possibility of its being used again remains.

The Factor of Experience in Layout Work. - The layout man cannot do a good job on the complete layout chart of the sign unless he is familiar with the materials used in its construction. Unless he is an accomplished glass blower himself, he may indicate bend backs and special glass formations very difficult to carry out in practice. It is well, therefore, to have a conference between the layout man and the tube bender before the final glass layout is decided. The same thing holds true of the metalwork and the electrical wiring. Eventually, of course, the layout man will become familiar enough with the work to be able to decide upon the complete layout on

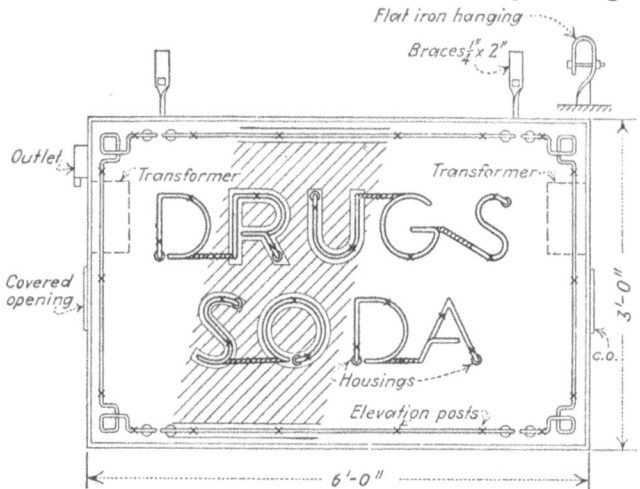

Fig. 39.—Complete sign layout, showing position of tubing, transformers, and details of metal drilling, etc. The specifications given in Table VII refer to this layout and the wiring diagram shown in Fig. 40.

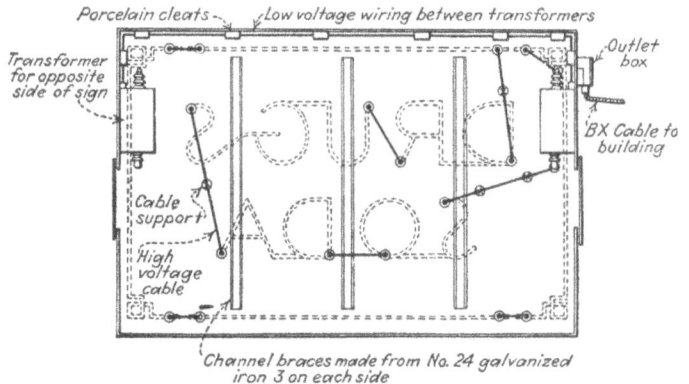

Fig. 40.—Wiring diagram of sign layout given in Fig. 39. Diagram shows one face of sign removed.

his own authority, and he can then be held responsible for the complete design of the sign.

The design principles given in this chapter are intended to guide only; the layout man may find that different procedures suit his particular plant much better than those given here. Nothing has been said about special design problems, such as obtaining special animated effects, high brilliance, unusual color combinations, etc. These are more advanced problems which are treated in later chapters.

70

Chapter Seven - Glass Bending

Equipment for the Glass Blower. - Good glass blowing is almost wholly a matter of practice and experience; it is an art which cannot be learned from reading a book. But the best glass blower is helpless unless he has the proper equipment with which to work, so it is well to discuss equipment first.

The requirements for a well-equipped shop include the following: a supply of illuminating gas under pressure; a supply of air under pressure; a ribbon burner; a cross-fire burner, or "cross fires"; and a hand torch. If pyrex glass is to be worked, a tank of oxygen gas and usually, also, a cannon type of burner must be used. But for ordinary lead glass, three types of fires the ribbon, the cross fire, and the torch will give the required flames to work the glass into whatever shape is required. In addition to this flame equipment, a perfectly flat table covered with a hard non-inflammable surface is necessary. A supply of heavy strong asbestos paper for making layouts and a supply of glass tubing of various diameters (see Table VIII) are the essential materials required. Chalk or china-marking crayon for marking the glass and heavy carbon paper for tracing layout patterns should also be available. A supply of corks to fit the various diameters of tubing is also required (see Table IX), together with various glassworking tools, such as a file or glass-cutting knife, glass tongs, a rubber tube for blowing purposes, and a gauge for measuring glass diameters.

Shop Layout. - The way in which the shop is laid out depends almost entirely on local conditions, but plenty of room should be allowed, particularly on the worktable, to take the largest piece of work to be handled by the shop. The straight glass tubing should be stored in racks near the table, often directly under it; the closer it is to the table the less time will be wasted walking to get it. The fires should be mounted on a separate bench about 3 feet from the long edge of the worktable and parallel to it. The flames should be mounted well apart so that a large piece of work will not get into two flames at once. A storage place for the asbestos layouts should be provided.

TABLE VIII.—CHART OF WEIGHTS AND DIMENSIONS FOR CORNING G-1 HEAVY-WALL CLEAR-GLASS NEON TUBING

Diameter, millimeters	Inches	Weight of 1 foot of tubing, pounds	Tubing per pound, feet	Number of 46-inch lengths per pound
9	0.3544	0.065	15.4	4
10	0.3937	0.073	13.7	3½
11	0.4307	0.078	12.8	3⅓
12	0.4724	0.090	11.1	3
13	0.5118	0.094	10.6	2¾
15	0.5905	0.122	8.2	2⅛

A convenient layout for the fire bench is shown in Fig. 41. From left to right the 18-inch ribbon burner, the four- or five-way cross fire, the 12-inch ribbon burner, the 6-inch ribbon burner, and the cannon fire are shown. The hand torches should be fed from rubber tubes, which make it possible to use them either on the fire bench or on the layout table. The layout table is low, not higher than 30 inches. The fire bench should be about 36 to 40 inches high. In many

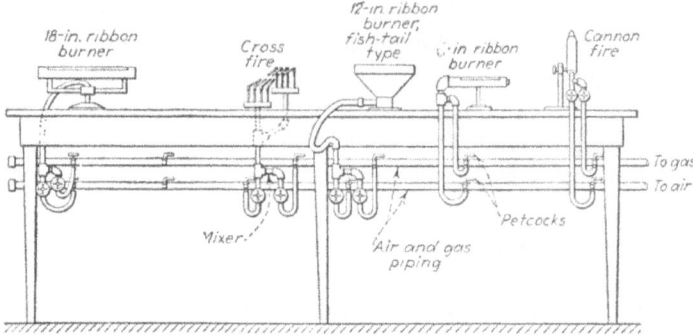

FIG. 41.—Typical fire-table layout, showing fires required for glass bending.

plants, the gas and air blowers are mounted directly under or to one end of the fire bench, so that the connecting pipes can be as short as possible.

The racks for the glass should be classified according to the diameter of the glass. Drawers for corks, electrodes, and tape should be provided in or near the layout table.

Buying Plant Equipment. - The mistake is often made by the man just entering the neon business of buying equipment which is just large enough to

TABLE IX.—CORK SIZE NUMBERS SUITABLE FOR PLUGGING TUBES OF VARIOUS DIAMETERS

Cork Size Number	Fits Tube of Outside Diameter, Millimeters
000	8
00	9
0	10 to 11
1	12 to 13
2	14
3	15
4	16
5	17 to 18
6	19 to 20

suit his immediate needs. A successful plant will grow very quickly in the first year, and plenty of room for expansion should be provided from the beginning. This fact is particularly important when buying air blowers and gas-boosting equipment. Extra fires can be added cheaply, but extra air- and gas-pressure facilities are expensive additions.

Some second-hand fire equipment is available on the market. This, given a fresh coating of paint and reconditioned, looks efficient and serviceable and in fact will give satisfactory service in most cases when first installed. But the buyer of second-hand equipment will find, as does the buyer of a second-hand car, that the air blower and gas boosters require continual repair and that the fires do not keep their adjustment. It is usually wiser, therefore, unless the buyer knows equipment values thoroughly, to buy new equipment from a reputable manufacturer, since this equipment will give first-hand service and is backed by a worth-while guarantee.

Gas Pressure. - In most cases, common manufactured illuminating gas is available to the shop owner. If natural gas or artificial gas in tanks must be used, the conditions are different, and page 132 should be consulted carefully. For ordinary gas, the pressure required at the burners is about ½ pound per square inch. Many gas supplies will give this pressure without boosting, provided that large pipes are used between the gas meter and the burners. In any event, pipe no smaller than 1½ to 2 inches in diameter should be used between the meter and the flexible tubing connected to the burners; if smaller pipe is used, it will lower the pressure to the point where a gas booster may be absolutely necessary.

In the case where sufficient pressure is not available, a gas booster must be used. The booster is a mechanical blower, driven by an electric motor, which takes the gas from the meter and boosts and maintains a constant pressure of the required value. Even when a booster is used, large pipes should be used to conserve the pressure.

FIG. 42.—Use of cross-fires. Note position of hands.

Air Pressure. - Air under pressure must be mixed with the gas so that the flame will have the required heat. The extra heat which can be obtained from gas by mixing air with it comes from the union of the oxygen in the air with the burning gas. The air pressure is obtained in most shops by a motor-driven blower. The blower should feed the air into a 1½- to 2-inch pipe, as in the case of the gas, in order to maintain an even pressure. About 2 pounds per square inch of air pressure is required.

The Air-gas Mixer. - An air-gas mixer should be used for each fire, as this is the best device for obtaining an even and uniform mixture of the gas with the air. A device known as a Venturi injector is combined with the mixer for better regulation. In ordering these devices, the number and size of the burners to be operated from each mixer should be specified. Mixers can be made from pipe couplings with either pet cocks or needle valves. Also, in some cases, mixers are made from glass. In all cases, the right mixer should be used for the job. The proper mixer mixes the air and gas in the right proportion and delivers this mixture at a constant rate to the burner.

Adjustment of the burner air and gas supply is accomplished by the use of pet cocks or needle valves in the gas and air lines as they enter the burner itself. In starting the burner, the gas pet cock is turned on first. Then, as the air is gradually admitted to the gas supply, the color of the gas flame will change from yellow through various shades of green and blue until the correct dark-blue flame appears. If too much air is admitted, the flame will make a roaring noise, and will assume a jagged appearance. Occasional readjustment of the pet cocks will be necessary, but the flames should remain steady for hours at a time without attention.

The Cross-fire Burner. - The burner shown in Fig. 42 is a cross-fire burner. It consists of five (sometimes six, three, or four) tubes on each side, each tube aimed so that the flame from it blends with the others, as shown. Each side of the burner may be provided with separate mixers, or the two sides may be brought together by piping to a common gas-and-air connection. The mixer connects the output of this connection with the burner itself. In practice, the burner is adjusted until the flame is a light blue cone, surrounded by a somewhat darker-blue glow. The outline of the central cone should be smooth, not jagged, and the flames from the opposite sides should be made to meet just midway between the two halves.

The cross fires are used for heating the glass tubing in a small area, that is, for making right-angle bends, other angle bends, and doubleback bends. In the hands of an expert, the cross fires also can be used for making gradual curved bends and splices, but these should be made in the ribbon burner and torch, respectively, for best results.

The Ribbon Burner. - The ribbon burner produces a solid ribbon of flame, as shown in Fig. 43. It is used for heating a long length of glass and softening the entire length uniformly. The gradual bends of the curved letters O, C, G, B, R, etc., are made in this way. The flame looks like a solid layer of light-blue flame, just over the opening of the burner, with a large mass of dark-blue flame extending for several inches above it. The width of the ribbon burner varies according to the kind of

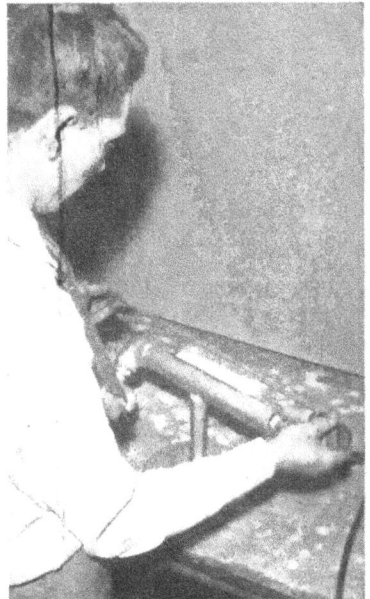

Fig. 43.—Use of ribbon fire. Note method of holding glass as it is rotated in flame.

work. For most shops, two or more burners of this type will be needed.

The Hand Torch. - The hand torch provides a small flame ranging from ½ to 1 inch wide and uses fishtail burners. The torch is held in the hand and applied to the work directly. It is used to heat a small section of glass completely and evenly around its circumference. It is used for splicing two pieces of tubing together, for sealing electrodes on, for sealing a tubulation and for sealing the work off the pumps, and as a general utility flame for awkward work. It also serves in most plants as the lighter for the larger burners and is often left burning continuously.

The Asbestos Layout. - Before the work of bending the tubing can be started, the working layout must be made. This layout is drawn in reverse, that is, with each letter backward, as the sign would appear in a mirror. This is done so that the glass can be laid down flat on the layout while it is being worked and thereby matched to the layout directly. The double backs, connections, and so forth will then appear at the back of the tube, where they should, rather than in the front, where they would be if a reversed layout were not used.

The layout is drawn on asbestos paper, since the hot glass may be pressed against such a material without fear of burning. Two methods are generally used in transferring the outlines of the letters from the original complete layout (see previous chapter) to the asbestos paper. The first uses a toothed wheel, which is rolled over the original layout, following the outline of the letters and producing a perforation or series of fine holes through the paper of the layout. The complete layout is then placed bottom side up on top of the asbestos paper, and powdered charcoal sprinkled over it. The charcoal leaks through the paper on to the asbestos and leaves the trace of the letters, in reverse. This trace can be strengthened with a charcoal pencil.

The other method of tracing the letter outlines on to the asbestos is the carbon-paper method, generally preferable to the perforation method. A piece of special black carbon paper, as large as the layout, is placed carbon side up on the worktable. It is covered with the asbestos paper, and finally the complete layout drawing is placed face up on top of that. Using a heavy pencil or pointed metal or glass stylus, the outline of the letters is traced over

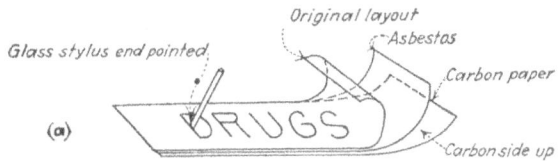

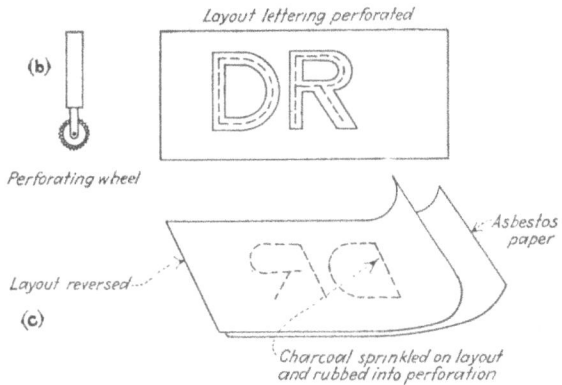

Fig. 44.—Preparation of asbestos layout. (a) Carbon paper method. (b) and (c) Use of perforating wheel and powdered charcoal.

on the original layout, and the carbon paper produces a reverse pattern on the asbestos paper. The pattern may be strengthened with charcoal pencil.

The glass blower's asbestos layout should be preserved after the sign is completed, as a record of the job, and it should never be used for another layout. Putting more than one layout on a sheet of asbestos is a common practice, but it is a false economy. Asbestos paper is cheap, compared with labor and other materials, and it should not be used too sparingly.

75

Glass Bending. - Each glass-bending operation involves first the softening of the glass in the proper flame. If the glass is thin or very thick, it should be warmed very gradually; otherwise, the severe and quick heating will crack it. Ordinary neon-tubing glass is very easy to work with and can be put directly into the flame without fear of cracking, but it must be kept in constant motion. This is the first requirement of glass blowing, that is, twisting or turning the glass in the hands while it is being heated, so that it will not melt in one spot before the entire surface has softened. This trick of turning the tubing becomes second nature after a little practice. The glass should be handled to

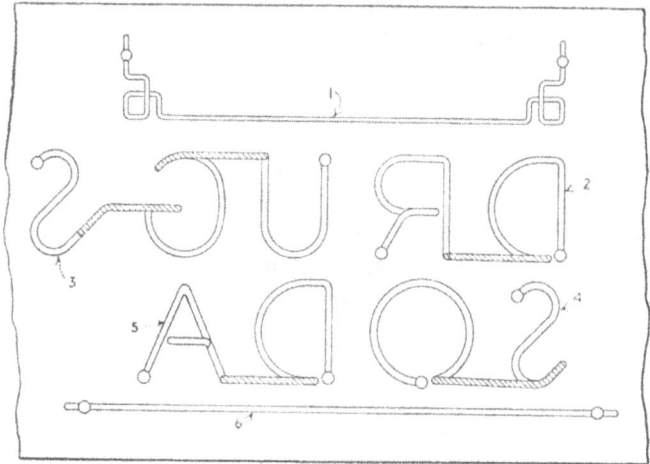

FIG. 45.—Asbestos layout of sign described ready for use by the glass blower. The specifications table given below should accompany the layout on a separate sheet. The numbers refer to the "unit" in specifications table.

SPECIFICATIONS

Unit	Number required	Kind of glass	Color	Type of electrode	Gas pressure	Mercury
1	4	15 mm. Noviol	Green	D-2	10 mm.	Yes
2	2	12 mm. Clear	Red	D-1	12 mm.	No
3	2	12 mm. Clear	Red	D-1	12 mm.	No
4	2	12 mm. Clear	Blue	D-2	10 mm.	Yes
5	2	12 mm. Clear	Blue	D-2	10 mm.	Yes
6	4	15 mm. Noviol	Green	D-2	10 mm.	Yes
Total feet						80 feet.

supply heat uniformly to the entire section being heated. When sufficiently hot, it will have a reddish glow, and it will flow like very thick molasses.

The glass used for neon-tube work, designated by the Corning company as G-1, has a wall thickness ranging from 0.045 to 0.060 inch. This thickness is greater than that of ordinary glass tubing of the same diameter, and as a result the neon type of glass can be treated more roughly both on the table and in the fires. This thickness of glass has been found to be, at least in the clear-glass variety, almost completely self-annealing. This means that it can be taken from the flame and set on the layout table without fear of cracking or internal weakness.

On no account should a piece of cold glass be allowed to come into contact with a piece of hot or melted glass. Both hot and cold glass will be cracked immediately. This point must be remembered particularly in the making of sharp bends, where the hot bend may accidentally hit another cooler section of the same piece of glass.

Marking the Glass Tube for Length. - Before the tube is put in the fire, however, it is first marked off for length. The tubing comes in 46-inch lengths. Although it can be procured in longer lengths if needed, the 46-inch length is a convenient one to handle. The cold tube is measured off against the layout, to see how many letters can be made from the length. Space for all bends and double backs must be allowed, and at least 6 inches on either end should be allowed to provide a cool end with which to handle the tube. When it is decided how many letters will be made from the one tube, the bending can begin.

The Angle Bend. - The angle bend, which may be a right angle or sharper or broader than a right angle, is a simple bend in the glass.

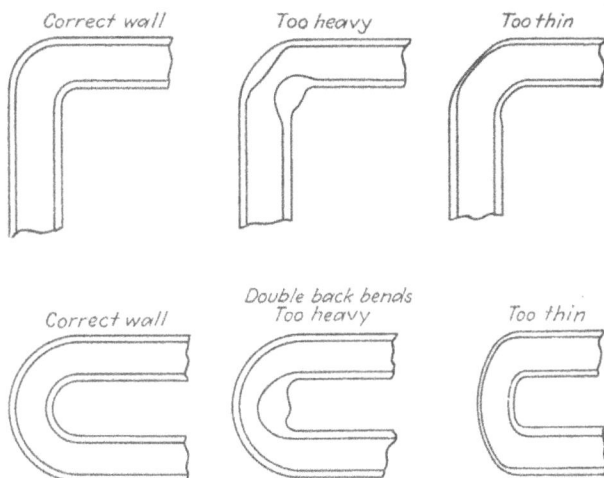

Fig. 46.—Correct and incorrect forms of angle and double-back bends.

The point to remember in all glass bending is to keep the thickness of the glass wall uniform throughout the tube, including all bends. If an angle bend were made without special care, the glass would become thick on the inside edge of the band and thin on the outside as shown in Fig. 46. To prevent this, the glass bender must see that the glass is rotated constantly in the fires and that the glass is melted evenly at all points. After it is heated and the bend made, the glass is blown out by the operator. One end of the tube is plugged with a cork (see Table XI) of the proper size to make it fairly airtight. The glass worker blows into the other end. The pressure thereby produced inside the tube

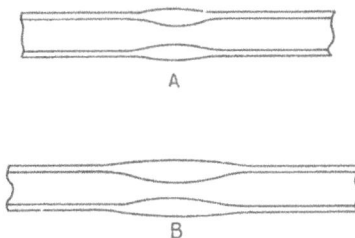

Fig. 47.—The glass must be gathered in the fire before a bend is made. (A) shows correct form for angle bend, (B) for a double-back.

forces the molten glass out into a uniformly thick wall, removes the strains, and produces a strong, neat bend. The knack of blowing the tubing, looking at it, and manipulating it almost at the same time requires consistent practice, and even accomplished glass blowers find that they need time to "work themselves in" after being inactive.

The cross fires are used to heat the glass for the angle bend. The glass should be "gathered," that is, pushed together after it has become slightly soft, so as to provide plenty of glass at the bend. If the angle is to be matched to an angle on the layout, the glass is blown and then laid on the asbestos while it is still soft. The angle can then be made to match the layout exactly. Occasionally it may be necessary to anneal the bend, that is, to cool it slowly in the fire, but usually this is unnecessary if regulation thick-walled neon tubing is used.

The Ribbon-burner Bend. - For the gradual curving or circular bends, the ribbon fire is used. The tubing, after being marked with chalk to show the position of the bend, is placed in the flame and rotated in it until the whole length of the tube becomes slightly soft. Not so much heat is required as for the angle bend. The tubing is laid against the layout while still soft and worked to match the layout. Blowing may be necessary, but the

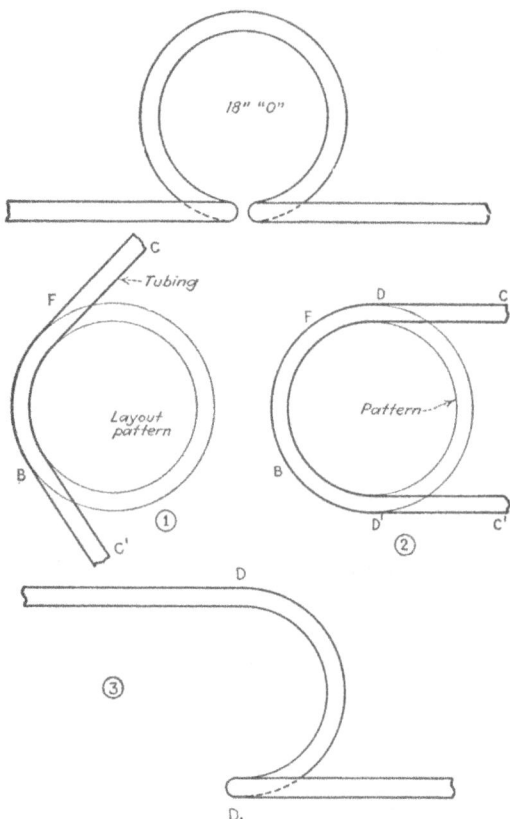

Fig. 48.—Various steps in making the letter "O," described in the text on the opposite page.

tubing should not get so soft that it begins to collapse anywhere, in this type of bend. If a large bend is to be made, it should be made in several steps, as shown in Fig. 48, always keeping the straight, unbent portion of the tube tangent to the curve on the layout.

For such letters as S, P, R, the circular bends are small - only half a letter in height. For C, G and O. they are larger and have to be made more carefully. The letter O, for example, in block lettering must be made perfectly round

78

and is therefore one of the hardest letters to bend. The following procedure for making a letter O, 18 inches in diameter, illustrates one of the most difficult bending operations. As shown in Fig. 48, two lengths of tubing 46 inches long are used.

1. Allow an 8-inch handle on the end for manipulation. Make the 12-inch bend *FB*, matching the arc of the circle, so that the ends of the tube *BC'* and *FC* extend from the circle on an exact tangent.

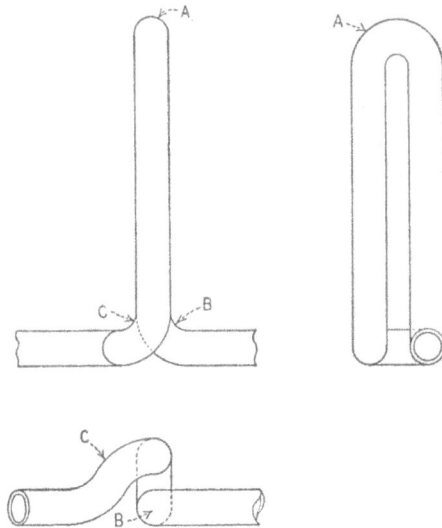

FIG. 49.—Bends used in making a double-back or "return." *A* is the double-back bend, *B* is a right-angle bend, and *C* is a combination bend.

2. Make the 18-inch bends *BD'* and *FD*, matching the outline of circle, so that the remainders *D'C'* and *DC* extend at a tangent as before.

3. Make another half circle, on the other length of tubing, using the same procedure.

4. Make a double-back bend (see below) in each half at *D'*.

5. Cut the two tubes at *D*, and splice together.

The Double-back. - For such letters as R, E, F, G, the tube must be bent back upon itself, as shown in Fig. 49-A. Such bends are known as double-backs. They are made in the cross fires, in somewhat the same fashion as the angle bend. A longer length of tube is heated, about 1 inch, and more glass is gathered together. When the glass is glowing red, the tube is bent sharply back on itself, as shown in Fig. 49, and blown carefully to preserve the proper wall thickness. When nearly stiff, the bend is completed as shown. When the bend back occurs at right angles to the face of the sign, as it does in almost every case, care should be taken to see that the rear part of the bend lies directly behind the front part; otherwise, the bend back will appear thicker than the rest of the tube, when the sign is lighted.

Combination Bends. - The type of bend shown in Fig. 49-C must be used in many cases, particularly in connection with the angle bend. This bend, sometimes called the combination bend, is in reality two bends close together. Its main purpose is to bring the rear piece of glass in a bend-back flush with the face of the sign and at the same time to make the required angle. The bend may be made in two steps, but usually the two parts are made at the same time. A sufficient length of glass for the entire bend is heated in the cross fires, and the glass manipulated and blown to give it the proper shape. Other bends which may be necessary in special work can usually be made up of combinations of the three fundamental types: the angle bend, the ribbon-burner bend, and the double-back.

Cutting and Splicing Glass. - Besides being proficient in bending the glass, the operator should be capable of cutting and splicing the glass properly before he begins to make a tube. Cutting glass appears to be simple, but even this operation requires a certain knack. The glass, marked with chalk at the proper point, is scratched with a file or glass cutters' knife. The file mark should be made at right angles to the length of the tube, and it should be a definite cut in to the glass about 1/8 inch long. The glass is then tapped gently with the file or knife, at a point exactly opposite

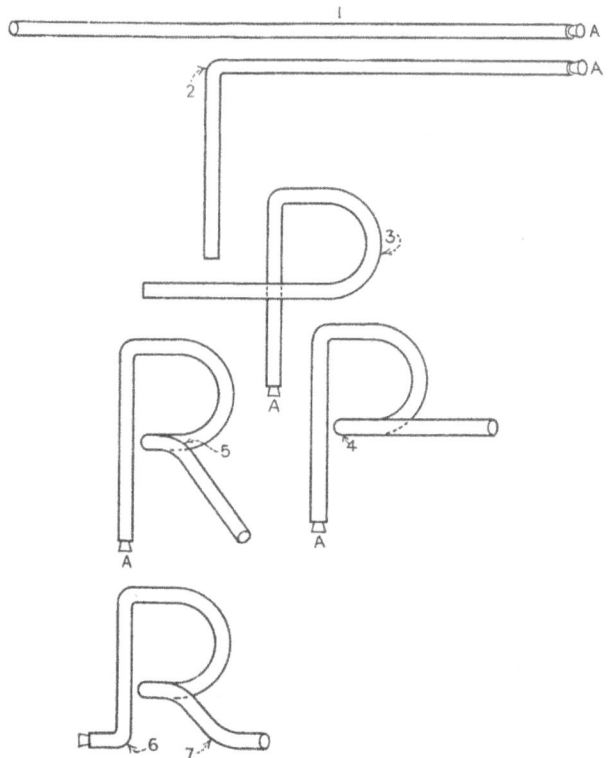

Fig. 50.—Steps in bending the letter "R."

the file cut, until it breaks. The break should be clean and even all around the circumference of the tube, as shown in Fig. 51 not uneven, as shown in the second picture of the same figure. Poor glass or glass which is not cool or not properly annealed will tend to break unevenly.

A most useful tool for making the initial file cut in the glass is a flat, hardened-steel file, 1/16-inch thick and ½ inch wide. The four edges of the file are used as cutting edges; as they become dull they can be renewed by running the narrow edge of the file along an emery wheel or dry stone.

For delicate work or when an especially clean cut must be made, the following procedure is recommended: File a hairline surface cut into the glass at the position of the required cut and at the required angle, which usually is exactly perpendicular to the bore of the glass. Then heat the end of another piece of glass tubing until it has melted into a small molten bead of glass. Press this bead against the file mark in the tube to be cut. The glass will part

80

at the cut in an even, smooth break. The experienced operator should be able to cut glass time after time with such success that every piece cut can be stood on its end without falling or wobbling.

When two sections are spliced, it is necessary to cork up one piece and to heat the two pieces of glass

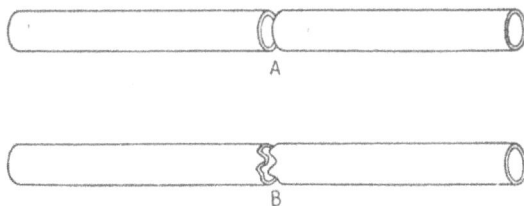

Fig. 51.—(A) A properly made cut in glass tubing. (B) A jagged cut, which cannot be used for a successful seal.

together in the flame either of the cross fire or the hand torch. In the cross fires, the glass is rotated; with a hand torch, the torch is rotated. When the glass has softened, the two edges are pushed together, and the open end of the tube is used for blowing the splice out to the proper wall thickness. The glass should be worked together and even reheated if necessary, to make sure that the two pieces are thoroughly fused together.

Splicing Large Pieces. - Splicing two sections of a tube together often involves two large arid unwieldy pieces of tubing which have already been formed into letters. When this is the case, the two sections should be laid on two asbestos-covered wooden blocks, on the worktable, so

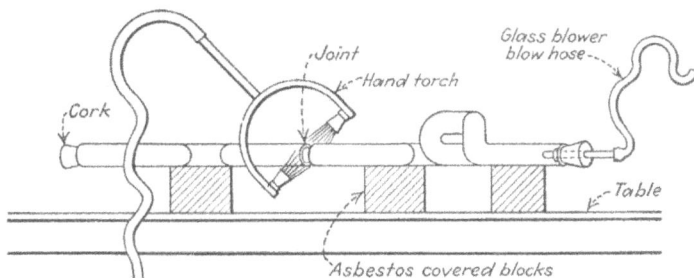

Fig. 52.—Method of splicing two sections of glass, showing use of asbestos blocks to raise the work from the table.

that the parts to be joined are firmly supported an inch or two above the surface of the worktable. The hand torch is used to heat the two ends of the tubing to be spliced, the torch being rotated around the glass, since the glass cannot be rotated in the flame. When the splice has been made but is not yet cold and hard, the blocks are removed, and the splice is laid over the proper part of the asbestos layout and made to match the outline of the layout, as in the case of the angle or circular bend. If it is impossible to move the tubing during the blowing process, whether it be a splice or a bend, a rubber-hose connection is used to provide a flexible means of blowing into the tubing. This tube should be about 5 feet long and is occasionally fitted at one end with a metal swivel joint, which prevents it from becoming twisted when the work is turned during the blowing process. It is important in splicing glass that only a small portion of each of the pieces to be joined be heated to the melting point. Usually, the glass should be heated not more than 1/16 inch

81

back from the end. A joint usually requires two heating periods, one before and one after the joining of the two pieces.

If the two pieces do not adhere completely, air will escape through one side during the blowing process. For this reason, the joint should be thoroughly closed before blowing is begun. If the joint is not successful in the first try, an expert may be able to repair it by reheating. But even the expert finds this job difficult; usually he starts over again rather than waste time on a hopeless task.

Splices in exposed portions of the tubing, such as in the middle of a sweeping curve, are conspicuous and should be avoided whenever possible. In large letters, however, such splices are often absolutely necessary, and in this case special care must be taken to make a neat job, with a uniform wall and no lumps or uneven patches.

Matching the Layout. - When the operations of bending, cutting, and splicing the tube have been mastered, work on a complete tube may be begun. The procedure is as follows: A mark is made with chalk about 6 inches from the end of the tube. This 6 inches is reserved for handling the tube. From this mark another length of from 2 to 6 inches is laid out for the turned-up portion which will be joined to the electrode.

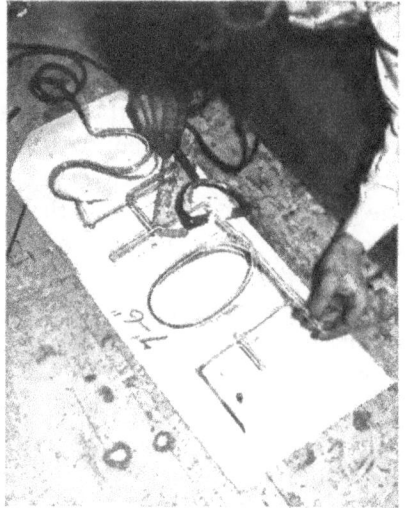

Fig. 53.—Glass blower splicing two letters with hand torch over asbestos layout.

The turn-up angle bend is not made, however, until all the letters have been formed. The second mark is the beginning of the first letter; this mark is laid against the layout at the beginning of the first letter, and the tube measured along the letter, around the curve if a circular bend is to be made or up to the first angle bend if an angle bend is to be made. A chalk mark is made on the glass to mark the location of the angle. Two marks should be made for every bend, the space between the two marks being the length of glass which will actually go into the bend.

The entire letter or group of letters is followed with the tube in this manner until all but the last 6 inches of the

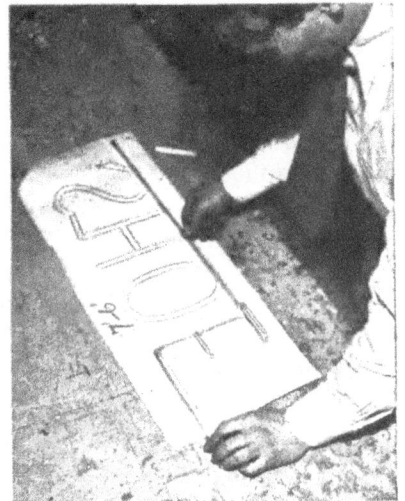

Fig. 54.—Matching letter to asbestos layout.

length of tubing have been marked off, these being reserved for handling and cutting. If this means that the useful length of the tube ends in the middle of the letter, it must be considered whether a splice in that place will be hard to make or will mar the appearance of the tube. If so, the last letter should be omitted, and the tube section cut off after the letter just preceding. When the length of tubing has been marked off in this manner, the bending begins. As each bend is made, the tube is matched to the layout while the bend is still soft, and the curves and angles are made to coincide exactly with the layout. In complicated letters like E, F, H, and so on, the middle double-back of each is made first, since this is by far the easiest way of doing the job. The bends on either side of the middle bend are then performed until the letter is complete.

For letters of 12 inches in height or larger, one or more 46-inch lengths of tube must be used for each letter, and splices must often be made in the

Fig. 55.—Matching a curved (ribbon-fire) bend to a layout.

middle of the letter. For smaller letters, often two or three letters may be made from one length of tube, the splices being made between letters where the painted-out portion appears.

The Influence of the Layout on Ease of Glass Blowing. - It can easily be seen that the layout has a great deal to do with the ease with which the tube is bent into letters. If unnecessary double-backs, awkward turns, or other errors of judgment are indicated in the layout, the glass blower's job is just that much more difficult. It is important, therefore, to use care and foresight in the layout of the pattern. The layout man may have a conference with the glass blower in order to decide just how the job can be done in the simplest and best manner. In general, the fewer the number of bends, provided that the letter outlines are still sharp, the better is the layout, and the longer a length of tubing can be made to go. Economy in the use of glass is a saving, but it should not be practiced if the glass blower's time is thereby increased, because the time is more expensive than the glass.

Attaching the Tubulation. - After the complete tubing unit has been formed, and before the electrodes are sealed on, an extra piece of glass, or *tubulation,* must be attached to the letter for pumping. The tubulation is placed at the rear of the glass, so that the seal-off will be hidden when it is mounted in place. The tubulation is attached as follows: A spot is chosen at the most convenient point for attaching the tubulation to the pumping mani-fold and to make the connection from the manifold to the tubulation short.

This point is heated with a hand torch, the heat being concentrated in a circle not larger than ¼ inch across. When the glass has become soft, the operator blows into one end of the letter, the other end being corked, and the glass at the heated spot blows out into a small bubble, which breaks arid leaves a small ring of glass to which the tubulation is spliced. The tubulation is simply a tube of glass, somewhat smaller in diameter than the tubing glass itself, usually 5 or 6 millimeters in diameter, and anywhere from 4 to 8 inches long.

It is spliced to the hole in the glass letter in the usual manner, the rubber hose being used for blowing, as in the splicing of two letters.

Since the entire pumping of the tube contents must take place through this tubulation, it is important that the path that it provides be wide and free. The splice should have the full diameter of the tubing, and in no case should a constriction be made, since it

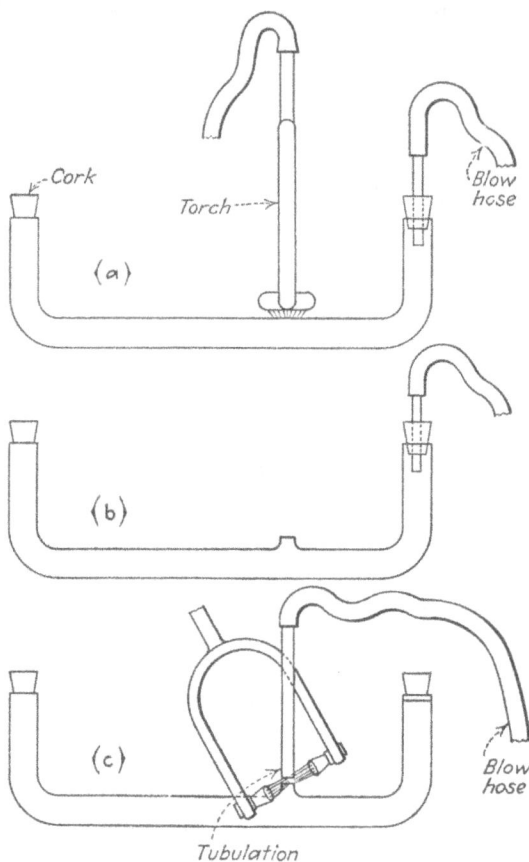

Fig. 56.—Sealing tubulation to glass letter. (a) Heating small area at back of letter. (b) Hole blown out. (c) Sealing tubulation glass to hole.

will greatly reduce the speed with which the tube can be pumped. Once the tubulation is in place, it may be used for all subsequent blowing operations.

Sealing on the Electrodes. - The electrodes which are fastened at each end of the tubing unit come provided with a glass jacket of approximately the correct cross section for a convenient splice. The larger electrodes may have a jacket whose diameter is too large for a good splice to the tubing unit. In this case, the jacket should be first reduced in size by heating the end to the melting point and allowing the glass to melt in. Once the electrode jacket and tubing have been brought to the same size, the splice is made in the same fashion as the ordinary splice between letters. One end of the tube is corked,

84

a rubber hose is attached to the tubulation, and one electrode is spliced. The cork is then removed, and the other electrode spliced in place.

Working with Colored Glass. - When colored lead glass is used, special precautions must be taken to cool it slowly, so that it becomes properly annealed. Otherwise, it will crack when subjected to strain. This is especially important when colored glass is joined to clear glass or to another type of colored glass. The general method of annealing, which should be followed when using colored glass, is called carbonizing and consists of holding the hot splice or bend in a flame from which all the air has been shut off. The gas of such a flame has a bright-yellow color, is not nearly so hot as the blue flame, and will coat cool objects held in it with a black deposit of carbon soot. When the splice is first held in such a flame, it is too hot to take this soot deposit;

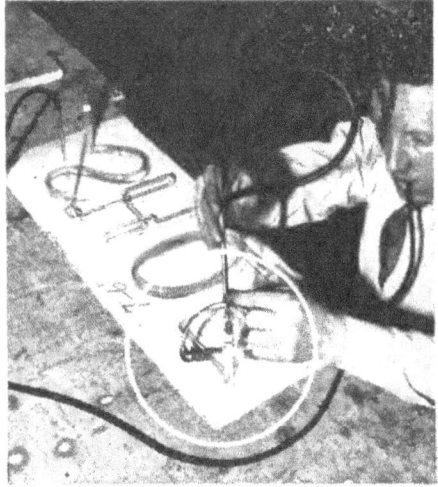

Fig. 57.—Sealing electrode to letter.

but after it has cooled in the flame sufficiently, the black deposit will form, showing that the annealing is complete. The annealing process is simply a slow and gradual cooling of the glass which eliminates the strains which would be set up in the glass if it were cooled in the air directly.

Working Pyrex Glass. - Pyrex glass has a much higher melting point than lead glass and requires, therefore, a much hotter flame for successful work. It is true that some operators find that they can work pyrex in the ordinary air-gas flame used for lead glass, but the work is slow, tedious, and rarely if ever perfectly done. The reason is that pyrex will just barely soften in the ordinary flame; and while it can be worked in this state, it quickly hardens and has to be continually reheated.

The flame used for working pyrex is a combination of the usual illuminating gas and oxygen obtained from a tank of compressed oxygen. Special burners are usually required to mix the gases and handle the extra heat. The cannon type of burner is particularly useful for work with pyrex.

The pyrex type of glass handles differently from lead glass. It can be put directly into the intense blue oxygen flame without fear of cracking. When hot enough, it works easily, and it will not discolor under intense heat. Furthermore, it requires no annealing, and it is very strong when cold. When molten, it glows with a white glow, as contrasted with the red glow of the lead glass.

In general, the same procedure of heating and bending is followed with pyrex as with lead glass, with the exceptions noted in the preceding paragraph. Pyrex glass cannot be sealed with lead glass or in fact with any other kind of glass except pyrex. [1] If such a splice is attempted, it will be found that it is impossible to fuse the two glasses together. When it is necessary for some reason to join pyrex to lead glass, a graded seal should be purchased and used for the job. This seal is a length of glass tubing, pale green in color, which contains seven different kinds of glass ranging from a lead glass at one end to a pyrex at the other. The lead-glass end is sealed to the lead glass, and the pyrex end to the pyrex.

If the seal to be made between pyrex and lead glass is on the pumping system or elsewhere where a solid joint is not needed, a ground-glass seal may be used. This is simply a mechanical fitted joint between the lead glass and the pyrex, very similar in outward appearance to the ground-glass joint in a stopcock. The joint is filled with a special vacuum-system grease which keeps it vacuum-tight.

Discoloration during Glass Bending. - If lead glass is overheated in bending or splicing, it will become discolored with a blackish deposit which forms in the glass itself. This color arises from the formation of free lead inside the glass, the lead being reduced from the lead oxide of the glass itself. The color may not be serious if it happens to occur in a place which will be painted out in the completed tube, but even in this case the glass is usually weakened. If the glass becomes discolored, it can be cleared up by heating the blackened part again in the fire and reducing the gas fed to the flame, leaving the air at the same pressure. Since the free lead has increased the melting point of the glass at that point, considerable heat may be necessary to make it flow, but usually the black color can be made to disappear. The glass should be cooled slowly from this point.

Cleaning Glass. - Glass which has dust or accumulations of other kinds of dirt and grease in it cannot be blown into a clean tube. Before work is started on a tube, therefore, it should be carefully examined. If it is clean, well and good. If not, it should be washed. Often water is sufficient to do the job, by a simple rinsing. But if stronger treatment is necessary, a solution of 2 or 3 per cent dilute hydrochloric acid (muriatic acid) may be used, followed by a thorough water rinse. Wet glass should not be put into the flame; it will crack at once.

Hints for the Glass Blower. - Even the best glass blower is not perfect. And glass has a way of being a very difficult material to deal with when things begin to go wrong. As an aid in such cases, the following list of hints is given, hints which if considered will aid greatly in obtaining good results.

1. Do not attempt to work hurriedly. A steady easy pace is best, and a surprising amount of work can be done if an apparently slow pace is kept up steadily. A good glass blower should be able to make about 60 feet of tubing in an eight-hour day. The rate of work depends a great deal on the kind of tubing bent, whether script or block letters, the size of letters, and tubing used; whether there are individual letters or a series of letters in one unit; and other factors.

2. Have the glass supply, electrode supply, tubulation glass, cutting tools, etc., close to the workbench. Walking for materials does not contribute to the bending of the tube, and it is an interruption which usually results in poorer work.

3. While bending the tube, keep in mind a mental picture of what the tube will look like when completed. This practice of seeing the bend in the mind's eye before making it will greatly aid the operator to do consistently good work without mistakes.

4. Insist on a proper air and gas supply and on burners which can be adjusted to give the proper flame. No single item in the glass-blowing art is more important than a correct flame, correct in both shape and temperature. Also, air and gas supply must be constant at all times to prevent continual readjustment of fires.

5. Guard against wasting glass. A good operator can save his employer much expense if he attempts to use glass economically. A glass waster is usually a poor glassworker.

6. If a splice or bend is bunched or lumpy, or if it is too thin, do not attempt to repair it. Cut back the glass and begin again; if absolutely necessary, begin the entire formation anew. This may not appear to be in line with the glass-conservation policy advised in the preceding paragraph, but a badly made splice can rarely be repaired no matter how much time is taken with it.

7. Do not attempt to make large curving bends too quickly. It is next to impossible to make a smooth curve if too much glass is heated at a time. Keep the unbent portion of the tube tangent to the curve on the layout.

Natural and Special Gas Supplies. - The manufactured gas available in most large cities is the lightest of the burning gases, and for this reason equipment designed for it cannot be used with other kinds of gas. In many sections of the country, especially near oil wells and coal mines, natural gas is provided, which is not so consistent either in burning quality or in pressure as is manufactured gas. Often this natural gas is pumped long distances, and it may be used in a city far removed from its source. For this reason, the purchaser of glass-blowing equipment should inquire of the local gas company as to just what kind of gas is supplied. If it is natural gas, burners should be ordered which are specially designed to use it.

If no gas supply of any kind is available in the city, gas in portable tanks may be purchased at a reasonable rate. This gas, known chemically as propane gas and sold under a variety of trade names, is the heaviest gas and can be used on the same burners as natural gas. Gas may also be obtained from gasoline, by bubbling air through the latter, but is rarely used for neon-tubing work.

Pressure Maintenance and Adjustment. - The pressure of gas supplies, especially of natural or tank gas, may vary from day to day, even from hour to hour. It is often wise to provide some method of maintaining a constant pressure. The most common way of maintaining gas pressure at a constant value is the use of a booster with a large storage tank. A diaphragm regulator may also be put in the line for further regulation, and still finer adjustment is possible if several tanks are placed on the line and fitted with a pressure gauge. A larger tank should be used for gas than for air; tanks as large as 50 gallons' capacity are sometimes used. The relief valve useful for air-pressure regulation cannot be used for gas.

A water manometer may be used for measuring gas pressure. The manometer is a U-tube made of plain glass of any diameter and half filled with water. One column of the U should be about 30 inches long; the other, about 60 inches long. The short column is tapped into the gas line; the other column is exposed directly to the air. Any difference in the level of the water in the two columns indicates a difference in pressure between the gas and atmospheric pressure. A difference of 1 pound is indicated by a difference in height of 27.8 inches.

For air-pressure measurement, a similar tube filled with mercury is used. Since 2 inches indicates a pressure difference of 1 pound, the long length of the tube need be only 12 inches long.

The Cost of Gas Supply. - Gas costs from five to eight cents per 100 cubic feet, depending on the quantity used, and, as can readily be computed, the cost for gas for a month in the glass-blowing plant is considerable, averaging from ten to twenty dollars per month. This cost can be reduced by keeping all burners off except when in use, but the practice is time consuming, as the burner must be lighted and adjusted each time that it is used. The other extreme of leaving the burners on all the time is usually very wasteful. The compromise usually used is known as an *economizer*. This is a valve, operated usually by a foot pedal, which allows just enough gas to get to the burner to keep it alight when not in use and which, when operated by foot, will admit the full supply to the burner. This device permits the greatest economy and is a great convenience, it is used principally on ribbon burners, since they consume the largest amount of gas.

[1] Nonex glass, one grade softer than pyrex, will seal to pyrex and is occasionally used.

Chapter Eight - Pumping Systems

The Pumping Problem. - The pumping system may well be called the center of the neon-tubing plant. The manufacture of tubing which will have good color and long life is virtually impossible if the pumping system is not up to standard.

The primary object of the pumping is to remove all the impure gas from inside the tube. Then, when the tube is filled with neon or whatever other rare gas is used, the gas in the tube will be pure and give a maximum light of the desired color. If even a small amount of unwanted gas is mixed with the rare gas, the color will be changed to a weak bluish light, and the sign will require a long aging period.

The second object of pumping is to remove impurities from the walls of the glass and from the electrode shells. The high-current gas discharge set up during bombardment heats up both the glass and the electrodes. This heat drives off the impurities from the walls and electrodes, in the form of gases and vapors. When the tube has become hot enough to scorch a piece of paper held against it; the pump proceeds to remove the gases and vapors which have been liberated from the walls of the tube and from the electrodes. If these gases are not removed very quickly, that is, if the pump does not work at a very high speed, the gases will be reabsorbed by the walls of the tube. Later, when the tube is installed and in operation, these reabsorbed gases may be liberated and destroy the brilliance and color of the tube's light. But if the pumping is rapid, the gases released by bombardment will be drawn out before they have had a chance to collect again on the walls of the tube.

After bombardment, the tube must be pumped out to a very good vacuum, that is, to a few millionths of atmospheric pressure. The rare gas is then inserted, and the tube sealed off the pumping system. The luminous tube is then complete, ready to be mounted on the sign.

The Requirements for the Pumping System. - From the foregoing it can be seen that the following requirements of a good pumping system must be met if good neon tubing is to be produced.

1. The pumping system must be rugged and absolutely vacuum-tight. The stopcocks used throughout the entire manifold and for the rare-gas containers should be of the high-vacuum type. They also must be specially treated with grease to make them free from even the slightest leak. Poorly ground stopcocks, no matter how well greased, will have a tendency to leak.

2. The pumping system must be clean. Dirt, water vapor, or excess stopcock grease in the system will greatly impair the vacuum.

3. The pumps should be capable of drawing a vacuum of 1 micron (about one-millionth of atmospheric pressure; see paragraph below, which explains the nature and meaning of vacuum measurement). This is a good vacuum, much higher than that used in electric light bulbs and about the same as that required in radio tubes. Mary pumps used by neon manufacturers today cannot produce so good a vacuum as this; but the tubing could be made more quickly and better if the pumps were capable of this performance.

4. The pumps should be fast. They should be able to draw a vacuum of 5 microns in 4 minutes. If they cannot do this, the gases freed during bombardment will not be pulled out before they are reabsorbed. Furthermore, long tubes, or tubes of large bore will require a great deal of time on the pumps, unless the pumping is fast.

5. The entire system should be reliable. A rough-andready system which will work regardless of the temperature or humidity and regardless of a certain amount of abuse is the only practical unit for a neon plant. Beyond occasional replacement of the oil in the mechanical pump and occasional replacement of the glass in the manifold, the pumps should require no servicing.

The Nature of a Vacuum. Before we go on to describe the various parts of the pumping system and their operation, it is necessary first to consider what a vacuum is, how it is measured, and how it can be obtained.

Three hundred years ago, man did not know how to obtain a vacuum. He may have imagined a space completely empty of every material thing, but he had never succeeded in actually producing a vacuum. Torricelli, the famous pupil of barometer, Galileo, discovered that if a glass tube, more than 30 inches long and closed at one end, were filled with mercury and then inverted into a glass also filled with mercury, as shown in Fig. 58, the mercury in the tube would be sustained to a height of not more than 30 inches. Above the mercury in the tube was, supposedly, a complete vacuum. We know now that in this space there is actually a quantity of mercury vapor, but at that time the existence of the vapor was not suspected. To this day, the vacuum produced in this way is known as a *Torricellian vacuum.*

It was soon found that the mercury was sustained to the height of 30 inches by the pressure of the air on the mercury in the cup. As the air pressure varied from day to day, the height of the mercury in the tube varied also, and it soon became known that the variations in atmospheric pressure were indications of changes in weather. The mercury tube thus became the first barometer. When the air contained much water vapor, its pressure increased, and the indication of humid weather was shown on the scale beside the mercury column.

Vacuum Measurement. - The relation between the air pressure and the height of the mercury column soon led to the use of the barometer as a measurer of pressures. In fact, most pressure gauges used in vacuum measurement today work on the barometer principle. It was found that the average pressure of the air at sea level was capable of sustaining a column of mercury to a height of 760 millimeters - 29.92 inches - which is the accepted standard of air pressure. In other words, it was convenient to express pressures in terms of the length (in millimeters) of the mercury column which that pressure would support. So today we find pressure measured in *millimeters of mercury,* abbreviated as "mm. Hg."

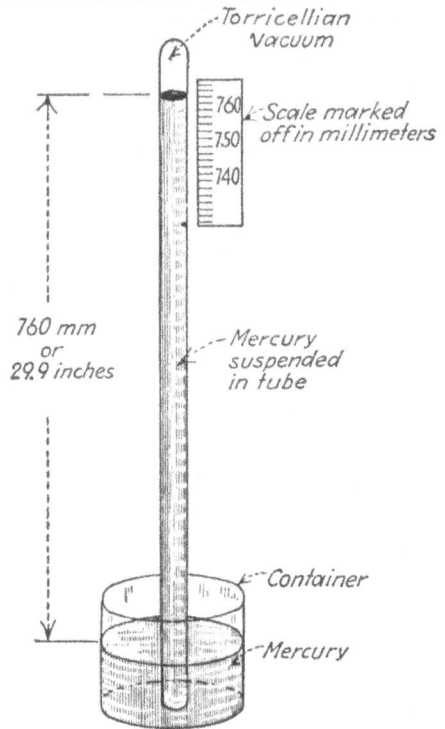

Fig. 58.—The mercury barometer, invented by Torricelli.

A complete vacuum is a completely empty space, that is, within which not 1 molecule of gas or other matter can be found. No perfect vacuum exists on earth, and even in the great empty spaces between the stars there are at least 28,000 molecules of matter in every cubic foot of space, that is, about 1 molecule to every cubic centimeter. The best vacuum ever obtained on earth by man was one in which there were 100,000,000 molecules in every cubic centimeter, and yet this apparently poor vacuum is equal to 0.000,000,00013 of atmospheric pressure.

Even with simple pumping systems, pressures of less than a millionth of atmospheric pressure can be obtained. Since such pressures would sustain a column of mercury to a height of only 0.007 millimeter, a more convenient unit for measuring vacuums, the *micron,* has been adopted. The micron is a thousandth of a millimeter. Good vacuums range from 0.01 micron at the very low pressures to about 5 microns at the high end. A good average vacuum for neon tubing before filling is about 1 to 5 microns.

Relation between Pressure and Vacuum. - The reason why the pressure inside a container is tied up with the degree of vacuum in the container is not always understood. A container full of gas is in reality full of millions of molecules of gas (as explained in Chap. Two). These molecules are flying about in every direction at very high speed. As the molecules near the wall of the container hit the wall, they exert a pressure on it. In other words, the pressure is really caused by a continual battering of moving molecules. Now, as the container is pumped out, fewer and fewer molecules remain, and consequently there are fewer molecules available to batter the wall of the container. In other words, the pressure goes down. Hence, as the vacuum increases, the pressure goes down, and thus the pressure can be used to measure the vacuum.

Pressure of Rare Gas. - After a neon tube has been bombarded and thoroughly pumped, it is filled with the rare gas (neon, helium, argon, etc.). The pressure of this gas is of the greatest importance. If the pressure is too low, the resistance of the tubing and the cathode-voltage drop will be high, and the tendency to sputter will be great. If the pressure is too high, the light will not be brilliant. So the pressure of gas inserted in the tube must be accurately measured. The same sort of gauge used for measuring the vacuum can be used to measure the gas pressure, except that the gas-pressure gauge must be able to read much higher pressures. The vacuum gauge reads microns, or thousandths of millimeters. The gas-pressure gauge reads millimeters (the pressure of neon gas inside tubes is usually between 5 and 15 millimeters, or roughly one-hundredth of atmospheric pressure).

One further fact about vacuums is of importance. When liquids are present in an evacuated vessel, they will tend to evaporate and fill the space of the container with a vapor. The pump will draw out these vapors along with the gases in the tube; but as long as any liquid remains in the tube, the vapor will be continually renewed. Hence the entire vacuum system should be free of liquids. If liquids, greases or oils must be used, they should be of specially chosen types of low vapor pressure, so that the amount of vapor that they give off in the vacuum system is small.

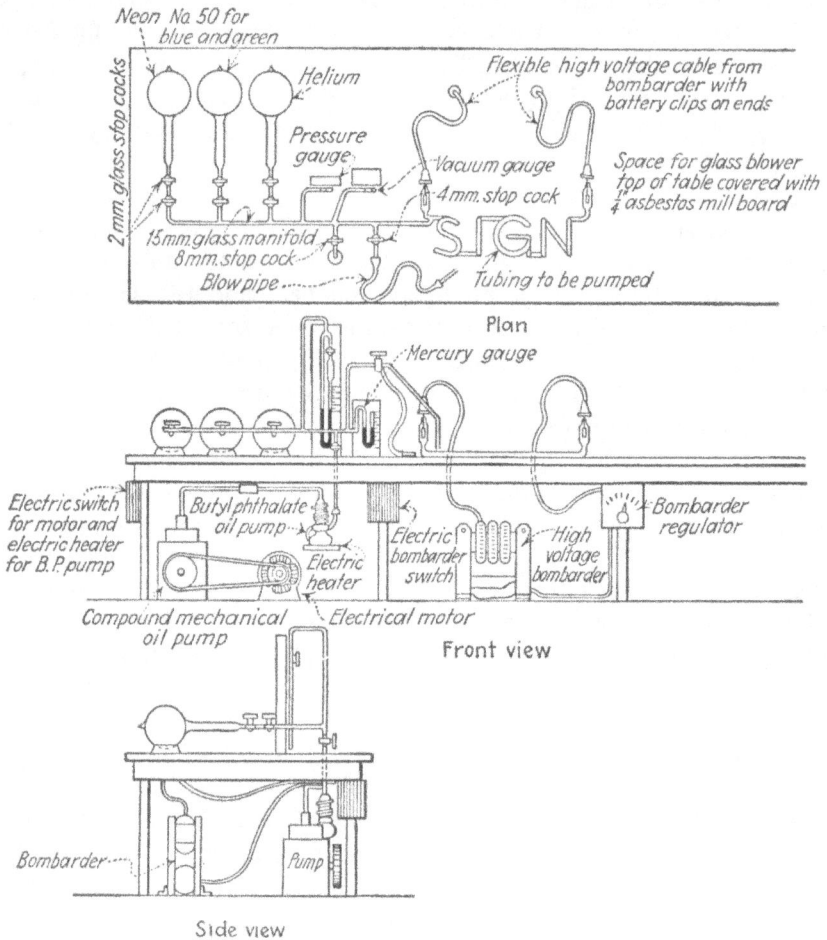

FIG. 59.—Complete layout of a typical pumping system for neon work. (The bombarding equipment shown should be compared with that shown in Fig. 75, page 173, which is of better design.)

The Parts of the Pumping System. - Pumping systems used by neon-tubing manufacturers vary according to the amount of work handled. Some systems are elaborate; others, very simple. But every pumping system must contain the following essentials: the pump itself, either a single mechanical pump or a mechanical pump and a diffusion pump in series; the pump manifold, with stopcocks; at least one vacuum gauge; and several containers of the rare gases, which are connected to the manifold through stopcocks. The more modern systems use two pumps, the mechanical pump being used to produce a rough *forevacuum,* without which the diffusion pump would not work. The diffusion pump does the actual work of producing the high vacuum.

92

Plant Layout. - The layout of the pumping system should be as compact and neat as possible. The one shown in Fig. 59 is a good typical example, but this particular style is not adapted to every plant.

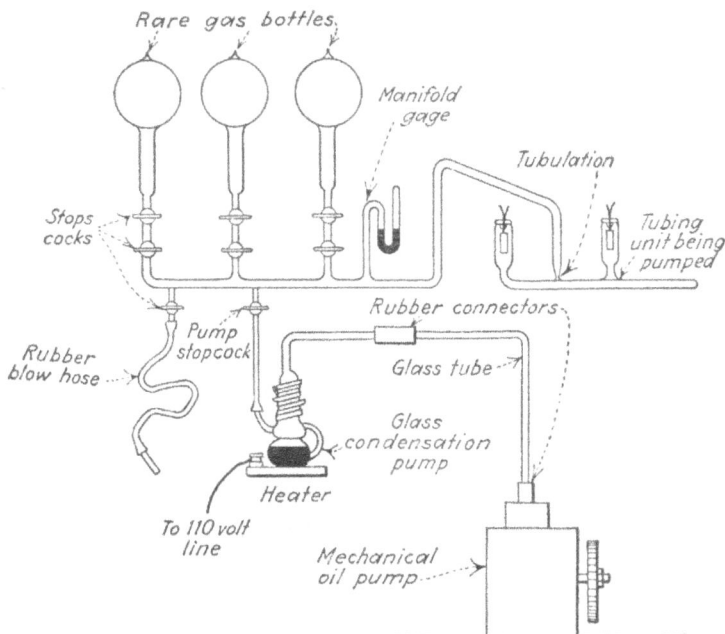

Fig. 60.—Schematic diagram of manifold system showing all connections.

Many pumping systems have been built haphazardly; gas containers and extra manifold sections have been added as the need arose. When a system has been built in this manner, it seldom presents a good appearance, and it is seldom as efficient as it might be. The pumping system should be built large enough in the first place to take care of the increase of work which can be expected as the business grows. The original layout should be given care and thought before it is actually installed. A neat layout is not a luxury. If the pump man works on a neat system, he will do better work. Repairs can be made quicker, leaks are easier to find, and the whole system can be kept in order with less bother. Of course, a pumping system which resembles that of a college laboratory is not necessary or desirable; but a good sensible layout is almost as important as are good pumps.

The Manifold System. - The pumping manifold is the part of the pumping system to which the tubes are attached for pumping. The manifold is a system of glass pipes arranged so that one or more tubes can be attached to it and having an assortment of stopcocks for use in connecting the pumps and the rare-gas flasks at the proper time. The glass used in the manifold system is usually the same as that used in the tubes, that is, lead glass. Pyrex is occasionally used, but since it will not seal to lead glass, it is not popular for manifold use unless pyrex tubes are to be pumped. The glass tubing used for man-

93

ifolds should be large - 15 to 18 millimeters in diameter - except where the tubulation is to be joined to the system, and, at this point, the tubing should be of the same size as the tubulation.

The manifold should be large enough to handle the work which will be put on it, but it should not be too long. If it is long, it will take a long time to pump out, and in addition it will waste the neon gas which must fill it when the tube is filled.

All splices in the glass manifold must be well sealed. Thin seals cause breakage due to vibration, and heavy seal? cause cracks due to strain and stress setup in glass. All glass should be well annealed.

A convenient manifold layout, showing three gas containers, all necessary stopcocks, pump connections, and tubulation glass for sealing on the tube, is shown in Fig. 60.

The rubber blowtube and stopcock shown on the manifold in Fig. 60 are used to permit the glass blower to blow into the system while sealing on the tube to be pumped and for the purpose of admitting air to the system whenever this may be necessary.

On some systems, special drying traps are used, which contain phosphorus pentoxide, a chemical which will absorb water vapor; if such a trap is used, the phosphorus pentoxide should be replaced regularly; otherwise it will become saturated with water vapor and do more harm than good.

Stopcocks. - The stopcocks used throughout the system are of standard design. They may be bought from glass-supply houses and are fitted with lengths of tubing at each end, which must be spliced to the rest of the system. The stopcock must have a carefully made ground-glass joint, as shown in Fig. 61. It is so constructed that when it is open, as shown in the figure, the turn handle is parallel to the length of the tube. It must be vacuum sealed, but even a good ground-glass fit is not sufficient unless specially treated with the *stopcock grease*, a substance somewhat resembling vaseline but specially made for vacuum work. This grease is smeared over the ground-glass

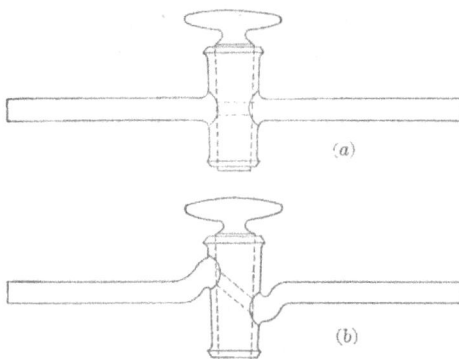

Fig. 61.—Two types of stopcocks: (a) Straight bore, (b) Oblique bore.

surface of the inner portion of the stopcock in a thin coat, but care must be taken not to get any grease in or near the hole through the stopcock, since this would clog the opening and allow the grease to get into the manifold. The inner portion (plug) of the stopcock, after being covered thinly with grease, is turned into the outside portion (or seat). When properly greased,

94

no air bubbles will be seen between the two ground-glass surfaces. When properly greased, the stopcock should turn easily and should be completely airtight. Grease used for stopcocks should have a very low vapor pressure; only the finest quality should be used, since poor-quality grease has a tendency to create impure vapors within the pumping system.

Stopcocks are made in two styles: with a straight and with an oblique bore, as shown in Fig. 61. The oblique bore has a great advantage in that it can be turned on in only one position, which makes it less apt to be opened accidentally. The bore of the main stopcock to the pumps should be at least 8 millimeters in diameter; otherwise, it will interfere with the pumping speed. Stopcocks have to be regreased occasionally, since the grease coat wears out after a time. Many hard-to-find leaks can be traced to neglected stopcocks.

The Rare-gas Containers. - The rare-gas containers must be attached to the system very carefully so that none of the gas is wasted and so that the gas inside the container remains pure. The method of doing this is outlined in Chap. Ten.

Some systems have only one stopcock on the rare-gas flasks. This simple installation will serve satisfactorily, but the double-stopcock installation is much safer, will conserve the gas, and will permit more accurate filling of the tubes, since after the first stopcock is opened, the second stopcock can then be opened to admit a small portion of the gas to the tube (see Chap. X on filling the tube).

Mechanical Pumps. - The mechanical, or motor-driven, pump can be used in either of two ways in the neon pumping system. It may be the only pump on the system; in this case, it must do the entire work of pumping, and for that reason it must be a very good pump. Or it may be used with a diffusion-type pump. In this case, the mechanical pump is used simply to provide a rough vacuum on the output side of the diffusion pump, while the diffusion pump does the actual high-vacuum pumping. In this case, the mechanical pump need not be capable of drawing so high a vacuum as when it is used alone. But it must be fast whether used with a diffusion pump or not. Regardless of its intended use, the mechanical pump should be carefully made to withstand long and hard usage.

The mechanical pump has two main outside connections: the intake port, which is connected to the manifold or diffusion pump; and the output port, which blows the pumped gases into the air. The pump can be connected to motor by a direct drive, chain drive, or belt drive. It should have the proper horsepower rating recommended by the pump manufacturer, and it should be firmly mounted on the same base as that of the pump itself.

The mechanical pump is filled with a special pumping oil which seals its valves and pumping action. This oil should be replaced occasionally, and it should never be allowed to fall below the level indicated on the inside of the pump casing. The oil becomes contaminated with water vapor and other impurities after about six months of use and should be replaced at the end of that time.

Rubber hose is usually used to connect the intake port of the pump to the glass tube which leads to the diffusion pump or manifold system. This hose should be of pure gum rubber and of extremely heavy construction (its wall thickness should be at least ¼ inch), and it should be bound to the intake port with wire. Sealing the rubber hose to the pump intake and to the glass should be done with castor oil or some other lubricant which will make the joint airtight.

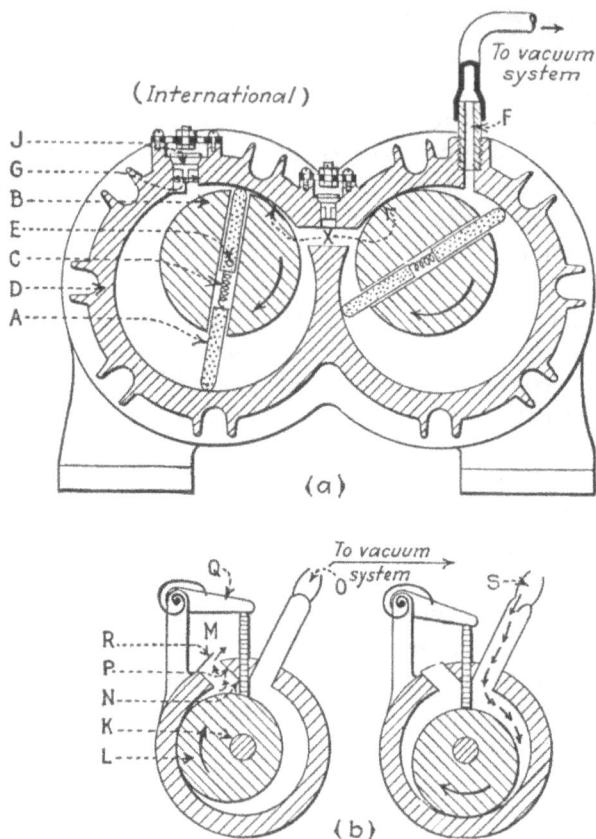

Fig. 62.—(a) The International compound rotary mechanical oil pump: A. Scraping vane. B. Rotor. C. Compression spring which forces vanes against cylinder wall. D. Cylinder wall (stator). F. Intake port. G. Outlet port. J. Compression valve. X. Rotor clearance.

(b) Cenco pump showing rotor in two positions. K. Main shaft. L. Revolving rotor. N. Scraping vane (fixed but movable vertically). O. Intake port. Q. Compression lever which forces vane against revolving cylinder. R. Outlet port. Two such units are operated in series in the Cenco pump.

Almost all modern high-speed vacuum pumps work on the principle shown in Fig. 62. The drive pulley drives a shaft on which a heavy vane slides, so that the edges of the vane are always in contact with the edge of the circular pump chamber. The drive shaft is off the center of the chamber as shown, so that the vane moves eccentrically. The vane, sealed against the side wall by the pump oil, sweeps the air admitted by the intake port around to the discharge port, where it escapes. This action is continually repeated as the pump revolves. Almost all pumps are really two pumps in one (compound pumps), connected in series and operated so that the two vanes turn together. This series arrangement adds greatly to the pumping speed and permits much higher vacuums to be attained.

The sweeping vane pump has been modified by one company into the eccentric cylinder type. The sweeping vane has been replaced by a cylinder mounted off center on the drive shaft. The cylinder rests against a sliding stationary vane which is sealed in the cylinder by the pump oil. As the cylinder rotates, it sweeps the gas from the intake port to the discharge port in much the same manner as that of the sweeping vane pump. In this pump, the sliding action is moved from the drive shaft to the side of the pump wall, and as a result the pump is easier to construct and maintain in good working order.

When new, the pump will draw a higher vacuum and will pump faster than after it has run awhile. Thus, while a new pump will often work satisfactorily without the help of a diffusion pump, after it has run for a period of time its ability to pump to a high vacuum will be greatly lessened.

TABLE X.—COMPARATIVE PUMPING SPEEDS

Type of glass pump	4000-cubic centimeter volume under laboratory conditions, minutes[1]		Estimate for 275-cubic centimeter average tubing units under plant conditions, minutes[2]	
	Hyvac forepump	Megavac forepump	Hyvac forepump	Megavac forepump
No diffusion pump........	33	20	7	4
Single-stage water-cooled mercury pump..........	18	16	Less than 4	Less than 3
Single-stage butyl phthalate pump................	20	..	Less than 4	Less than 3
Three-stage water-cooled mercury pump..........	9	7	3	2

[1] The figures given represent the time required to pump out the volume indicated to a limiting pressure of $\frac{1}{2}$ micron. P. J. Mills, "Time Pressure Characteristics of Various Diffusion and Molecular Pumps," *Review of Scientific Instruments*, vol. 3, No. 6. June, 1932.

[2] Authors' estimate.

Other types of pumps are occasionally used, one of which works on a rotating-gear principle, the teeth of the interleaved gears being used to sweep the air from intake to output, but these pumps quickly wear after a few months of service. In addition to the gear pump, there are piston pumps very similar in action to the piston hand pump used for filling automobile tires but with the action reversed. Piston pumps are not fast, and they cannot produce a high vacuum, so they are rarely used.

The most effective type of mechanical pump is the so-called *molecular pump.* This works with a reduced pressure vacuum on its output side, provided by an auxiliary pump. The molecular pump works on the principle of molecular drag; that is, a rotating vane is used to drag, rather than to push, molecules from the intake to the output. When properly operated, molecular pumps can give vacuums of 0.0004 micron when provided with a backing pressure of 1 micron. But they are expensive and delicate pumps, and they are not used in the neon trade.

The oil in all mechanical pumps will be sucked up into the manifold or diffusion pump as soon as the pump is turned off unless special precautions are taken. Some mechanical pumps are fitted with a special nonreturn valve which prevents the oil from backing up out of the pump. Other pumps do not have this arrangement. No matter what kind of pump is used, it is always best to allow air to enter it from the manifold after it is stopped.

Occasionally even this precaution is of no avail, and the pump oil does get into the pumps and manifold. When this happens, there is nothing to do but take down the entire system; clean it out with lye, carbon tetrachloride, or some other grease remover; refill the diffusion pump with mercury or butyl phthalate; and seal up the system.

Table X gives the pumping speed and highest attainable vacuum of two of the better-known kinds of pumps.

Diffusion Pumps. - Very few neon plants today depend upon a mechanical pump alone. Almost every neon manufacturer has come to realize that some sort of diffusion pump is necessary for a high-speed pumping system. The reasons for the superiority of the diffusion pump are many: It is independent of humidity changes, pumping equally well in wet and dry weather. It is extremely fast. It does not wear out, and it requires a back pressure which can be supplied by any fast mechanical pump.

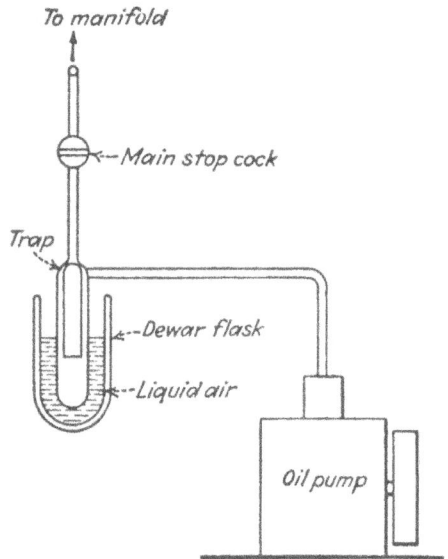

Fig. 62-A.—Simple mechanical pump with liquid-air moisture trap.

There are two types of diffusion pump commonly used: the mercury type and the butyl phthalate (bew-til tha-late) type. The names refer to the liquid which is used in the diffusion pump. These pumps consist of a glass reservoir containing the liquid, a heater which heats the liquid, a cooling chamber in which the vapor from the liquid is recondensed into liquid again by means of

98

a special cooling system, and a tube for conducting the recondensed vapor back into the main liquid reservoir.

The layout of a mercury diffusion pump is shown in Fig. 63. In the mercury pump, the pool of mercury is heated by the heater, and the vapor which forms above the pool is drawn toward the intake part of the mechanical pump. But the vapor

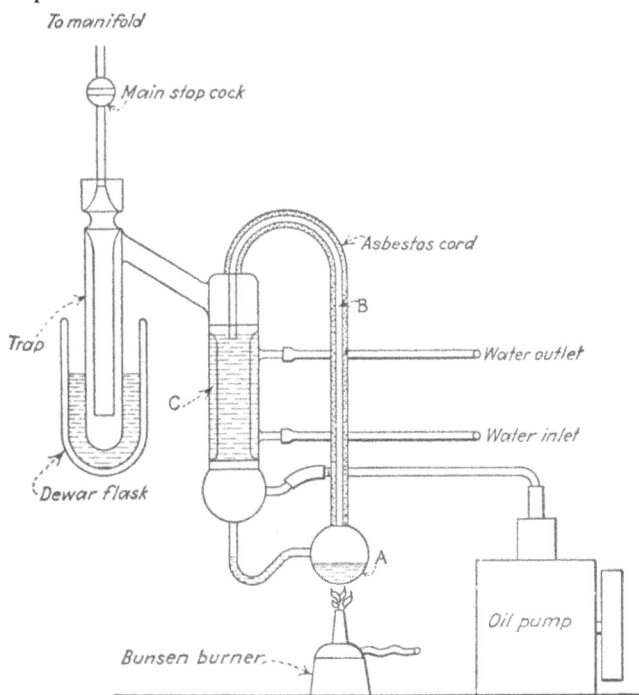

To manifold

Main stop cock

Asbestos cord

B

Water outlet

Water inlet

Trap

C

Dewar flask

A

Oil pump

Bunsen burner

FIG. 63.—Water-cooled mercury condensation pump. A. Mercury reservoir. B. Asbestos covered vapor chamber. C. Cooling chamber.

chamber is also connected to the glass manifold, and as the cloud of mercury vapor is drawn into the mechanical pump it sweeps before it a portion of gas from the manifold. The mercury vapor is the pumping agent; it half drags, half pushes before it the gases from the manifold into the mechanical pump.

If the mercury vapor were not collected, it would pass out into the mechanical pump and be lost, and eventually the supply of mercury in the pool would be all evaporated. So some means of collecting the vapor must be provided. This takes the form of a water jacket which cools the vapor chamber. By the time the vapor has reached the end of the chamber, this cooling effect has condensed it into tiny droplets which collect in the side tube and roll back into the main reservoir. By the time the condensation takes place, a fresh supply of vapor has formed and is pushing more gas from the manifold into the mechanical pump, so that the action is continuous as long as the heat is supplied to vaporize the mercury and the water cooling to condense the vapor.

A good mercury pump requires a cumbersome water supply to give the proper cooling. In addition, the presence of the mercury vapor in the system provides many difficulties if any other kind of tube but mercury is being made. For a straight neon tube, for example, the mercury vapor from the diffusion pump may get into the tube and cause a blue glow where only red is

99

wanted. So a specially devised vapor trap must be provided between the diffusion pump and the manifold. This trap, shown in Fig. 63, is a tube surrounded by a very cold substance, such as liquid air or chopped-up dry ice. The intense cold at this tube condenses any mercury vapor and traps it before it can reach the manifold system.

Only the very best and purest grade of mercury should be used in mercury pumps. Triple-distilled U.S.P. mercury can now be obtained at reasonable prices from a variety of sources, and no other mercury should be used for the purpose. After about six months of continuous use in a neon plant, on the average, the mercury in the pump must be replaced. During its use, it picks up a great deal of dirt, moisture, and other impurities which are drawn into it from the manifold. These impurities make the surface of mercury thick and viscous, so that eventually it does not boil readily; when this happens, the pump should be removed and cleaned and then refilled with fresh mercury. The cleaning of the glass in the pump can be accomplished by successive washings with nitric acid and hot water, a good general procedure for cleaning any part of the glass in the system containing mercury. The mercury can be distilled by the neon owner manufacturer if he wishes to go to the trouble of making a still. But since the amount of mercury used in the pump is small, and replacements are required only twice a year, very few pump owners distill their own mercury.

The Effect of Temperature on the Operation of the Pump. It is of greatest importance to maintain an even temperature under the mercury boiler of the pump. If the temperature is too high, the pressure of the mercury vapor will be too high, and the vapor will not recondense thoroughly before it gets to the mechanical pump. If the mercury gets into the bearings of the mechanical pump, trouble will surely follow, and usually the pump will have to be overhauled or replaced. Too high a temperature also impairs the effectiveness of the pumping action, since the greatest amount of gas will be drawn from the manifold only if the vapor pressure is correct not too high or too low. If the temperature is too low, not enough vapor will be produced to produce a heavy drag on the gas in the manifold, and the pumping will be slow. The proper temperature is that at which the mercury first begins to show definite signs of boiling. Violent boiling may wreck the tube; and even if the boiling is moderate, the "bumping" of the mercury may cause trouble. For this reason, a wide shallow mercury pool should be used in the boiler, rather than a narrow deep one.

The mercury pump, as can be seen, requires many accessories. It must have a complete water supply, with piping, rubber tubing, and valves for its control. And before pumping can be started, the vapor trap must be supplied with liquid air, which is very difficult to handle, or with dry ice mixed with acetone, which is also messy to work with. Nevertheless, this type of pump can supply the highest and fastest vacuum of any simple and commercially available pump, and it is very popular.

There are several other names for the diffusion type of pump such as aspirator pump and condensation pump, but the pump is the same regardless of which name is used.

The Butyl Phthalate Pump. - Recently a substance (butyl phthalate) has been discovered which will take the place of mercury in the diffusion pump. This liquid is for all practical purposes just as effective as mercury, and it requires no water cooling and no vapor trap. The principle of operation of the butyl pump is exactly the same as that of the mercury pump, but, since the vapor will recondense much more readily, less elaborate cooling measures are required. In fact, as shown in Fig. 64, a butyl

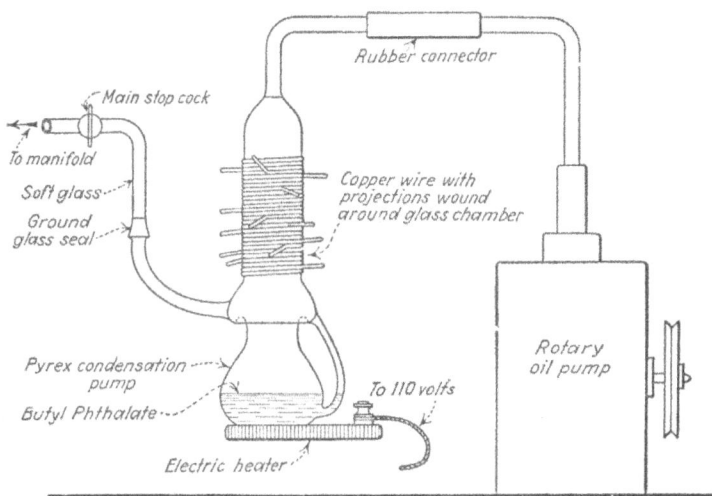

Fig. 64.—Air-cooled butyl phthalate condensation pump.

pump requires for cooling only a simple coil of copper wire. This coil radiates the heat from the glass wall of the vapor chamber and in doing so cools the chamber sufficiently to condense the butyl vapor.

The vapor of this liquid has a very low pressure; that is, at ordinary room temperatures very little vapor is given off, and hence the pressure is very low. In fact, the pressure of butyl vapor in the tube being pumped is so low that it causes no trouble. For this reason, the butyl pump requires no vapor trap and hence no liquid air or dry ice. The butyl pump is a very simple device to install and operate compared with the mercury pump, and it is gaining great favor in the neon trade. In fact, it is really nothing more than a device connected between the manifold and the mechanical pump which requires a proper electric heater and nothing else. It requires no attention except to see that the proper temperature is maintained by the heater, and its life is indefinitely long.

There are many designs of butyl phthalate pumps including a unit with an alcohol cooling jacket. Also butyl phthalate pumps are built having two or three pumping stages in one glass unit.

The butyl phthalate which must be poured into the pump just before it is sealed into the system must be kept pure and covered before using. The substance contains an impurity known as phthalic anhydride, which cannot be readily separated from the liquid. So long as the anhydride remains in the pump, the pump will not operate properly. The best procedure is to close the main-pump stopcock and to start the mechanical pump. Bringing the butyl phthalate up to the boiling point with the heater will gradually remove the anhydride. This procedure must be carried out before the pump is used, since it will not operate successfully if the anhydride is present. Even the very best and purest grades of butyl phthalate contain this impurity, and it must be removed before use.

There are one or two other substances which will serve in place of butyl phthalate. Butyl benzyl phthalate and the Apiezon oils can be used, but they are very rare and expensive and are used only in physical laboratories where the conditions are extremely exacting.

Temperature Regulation. - The temperature of the butyl phthalate must be regulated as carefully as that of the mercury in the mercury pump. The electric heating element used can be made of a horseshoe-shaped piece of asbestos board wound with 10 or 15 feet of nichrome resistance wire. If the heater provides more than 60 watts, it should be connected in series with the proper incandescent lamp, on a 110-volt circuit. This heater will provide the proper heat for the butyl pump; but it will not boil the liquid. If the butyl phthalate boils, it is being overheated.

Butyl pumps require a lower backing pressure than the mercury type of pump; they operate at maximum speed when the backing pressure is about 0.01 millimeter of mercury. Almost any good mechanical pump is capable of providing this back pressure for the pump, so that manufacturers need have no fear about replacing their mercury pumps with a butyl pump. Butyl pumps will not reach so low a pressure as will the mercury type, but the pressures obtainable with the butyl type are far below the minimum required for neon work. In general, it may be said that the mercury and butyl pumps will perform equally well for use in neon plants, but the greater convenience of the butyl type and its ruggedness make it preferable to the mercury type for this kind of work.

Diffusion pumps are usually made of pyrex glass, since this glass can stand heating much better than can lead glass. When such a pyrex pump is connected to the lead-glass manifold, either a graded glass seal or pyrex-tolead-glass ground joint must be used to connect the pyrex to the lead glass, as explained earlier.

The heater used is usually of the electric type, since this is convenient and comparatively safe compared to a gas flame. But gas can be used just as well, provided that too intense a flame is not used. When mercury is used, the heat should be just enough to make the mercury bubble gently as it boils, while in the butyl pump no visible boiling should occur.

The diffusion type of pump, of either the mercury or the butyl type, is usually bought in complete form and ready to install by the pump man when he builds the pump system. An accomplished glass blower can make such a pump, but it usually is cheaper to buy the unit complete.

Vacuum Gauges. - Besides the manifold and pumps already described, the vacuum system must have one or more vacuum gauges for measuring the degree of vacuum in the tube. All vacuum gauges used in neon work operate on the barometer principle; that is, they are really pressure-measuring devices. The simplest type of vacuum gauge resembles the mercury barometer, as shown in Fig. 65. This type of gauge, known as the "straight gauge," consists of a straight tube at least 30 inches long and filled with mercury. The tube is fastened in an upright position over the mercury reservoir and is sealed into the manifold, as shown. When the manifold is full of air, all the mercury is in the cup; but when pumping is started, the mercury column rises in the tube about 30 inches. The height of the column below the zero level is an indication of the degree of vacuum. The tube is calibrated with a scale marked off in millimeters, which is fastened beside

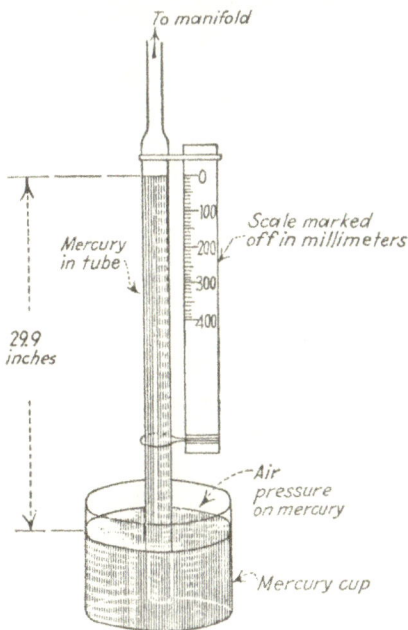

Fig. 65. — Straight gauge.

the tube. When the pressure falls to 10 millimeters, the column is only 10 millimeters below the zero level. This gauge indicates only a rough vacuum and is used to measure gas pressure only.

When the neon or other rare gas is put in the tube, its pressure may be measured with such a gauge. To secure a 10-millimeter pressure, the mercury column should fall 10 millimeters; for a 15-millimeter fill, the column will fall 15 millimeters; and so on. For very low pressures of 1 millimeter or below, this type of gauge cannot be used, because the length of the mercury column is so long; and its changes in length are so small, that the pressure cannot be read accurately even with a magnifying glass. In other words, the straight gauge is a *millimeter gauge;* that is, it will read pressures from 1 to 760 millimeters but not below 1 millimeter. For reading microns, or thousandths of millimeters, a micron gauge is needed. The McLeod gauge, a micron gauge, is described in a later paragraph.

The U Gauge. - The U type of vacuum gauge, shown in Fig. 66, is a millimeter gauge also, but it is much smaller and more convenient to install than the straight gauge. The U gauge is a U tube half filled with mercury and sealed into the manifold system. One arm of the U tube is evacuated. When the system is pumped out, the pressure on the manifold side of the gauge is reduced, and the mercury in the two arms of the U tube tends to come to the same level. The difference in level between the two arms, in millimeters, is the pressure on the manifold side. The scale of such a gauge is placed beside

103

the evacuated arm. The mercury in this arm will fall only ½ millimeter for every millimeter's decrease in pressure in the manifold. This is true because the other arm rises ½ millimeter at the same time, making the difference in level 1 millimeter. Since the U-type gauge moves only half as far as the straight gauge, for the same decrease in pressure, it is less sensitive and harder to read, particularly for pressures in the region of 1 millimeter. But because of its simplicity and great usefulness as a vacuum indicator, almost all pumping systems are fitted with a U gauge. The U gauge will show very quickly just what the pressure is inside the manifold, and it will show also how quickly the pumps are working and if there is a leak in the system.

The mercury used in the straight and U-type gauges does not interfere with the drawing of a good vacuum, because the mercury is cold, and, as a result, practically no mercury vapor is formed. For this reason, no special trap is necessary, as is necessary in the case of the mercury diffusion pump. Tubing used in gauges should be free from water vapor when the gauge is installed.

Fig. 66.—Standard mercury U gauge for measuring pressures from 1 to 30 mm.

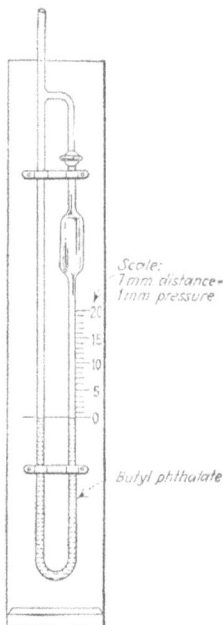

The Butyl U Gauge. - A new type of U gauge which is very useful in reading pressures from about ½ millimeter to 30 millimeters uses butyl phthalate instead of mercury. Since butyl phthalate is about one-fourteenth as heavy as mercury, the pressure inside the manifold will support a column of it fourteen times as high as it will support mercury; that is, a change in pressure of 1 millimeter in the manifold will cause a change of 14 millimeters in the height of the butyl phthalate column.

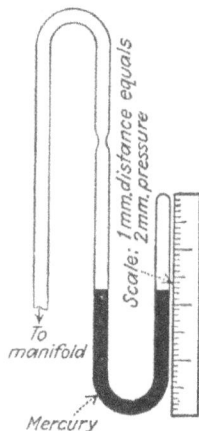

Fig. 67.   Fig. 68.

Gauges for measuring rare-gas pressure with great accuracy. Fig. 67.—Butyl phthalate U gauge. Fig. 68.—Dubrovin gauge.

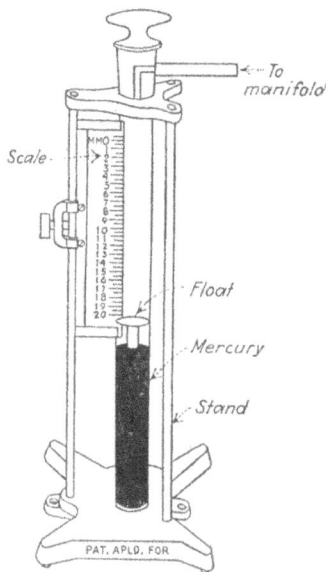

104

As shown in Fig. 67, this fact makes the scale of the gauge much wider and easier to read accurately. Filling to pressures as low as 1 millimeter and filling to higher pressures to an accuracy of 0.1 millimeter is entirely possible with such a gauge, while with mercury U gauges such accuracy cannot be attained.

The butyl phthalate used here is the same substance as that used in the butyl diffusion pump, and it has the same advantages of low vapor pressure and chemical inactivity. The complete butyl gauge is about 2 feet long and fitted with a scale about 8 in. in length.

The Dubrovin Gauge. - Another type of gauge having an extended scale is the Dubrovin gauge. This operates on the principle of buoyancy. A closed evacuated cylinder, weighted with lead shot, floats in a pool of mercury, in a vertical position so that the top of the cylinder lies opposite a scale. When the pressure on the mercury pool is low, the cylinder will float high and thus indicate the high vacuum, while, if the pressure is high, the cylinder will sink lower! By properly proportioning the weights, the scale of the Dubrovin gauge can be magnified many times over the standard millimeter scale. The operation of the device is illustrated in Fig. 68.

The McLeod Gauge. - The McLeod gauge is primarily a micron gauge; that is, it is designed to read very low pressures, as low as 0.1 micron. But the McLeod principle is also used in gauges designed for reading millimeters of pressure, and two gauges can be built into one, capable of both readings high and low vacuums. In Fig. 69-B, a double McLeod gauge of this type is shown.

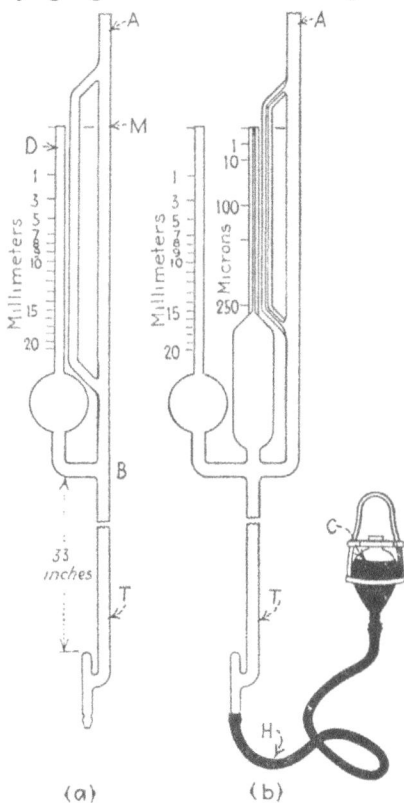

FIG. 69.—Two forms of McLeod gauge, described in the text, (a) is a simple millimeter gauge, while (b) is a double gauge for reading both millimeter pressures and micron pressures.

The McLeod gauge operates by compressing a sample of the gas from the manifold and comparing the pressure of this compressed sample with the manifold pressure itself. In Fig. 69, the cup C holds a supply of pure, clean mercury, while the rubber hose H connects the mercury reservoir with the gauge itself. The gauge is connected to the manifold at the point A. It is operated by lifting the mercury flask above the bottom of the gauge. The mercury from the flask flows through the hose and up the tube T to point B. When it reaches this point, it divides into two paths. The left-hand passage (which

105

contains gas at the manifold pressure) is cut off from the rest of the system by the mercury. As the mercury continues to rise in the tube D, the gas caught inside it is compressed. At the same time, the mercury in the right-hand tube rises. When the mercury in this column has risen to the mark M, the level in the closed chamber D comes to rest opposite a number of the scale which shows the pressure of the gas inside the manifold.

The McLeod gauge does not read continuously; that is, each time a reading is desired, the mercury flask must be lifted, and the reading taken after the mercury has risen to the proper height. The pressure reading indicates the pressure at the time when the mercury passed the point B, that is, when the sample in the chamber D was caught and the compression began. Reading pressures with this type of gauge is slow, therefore, and not adapted to rough work. In addition, the McLeod gauge cannot be used for measuring the pressure of the gas fill except by a cut-and-try method, which is slow and often very wasteful of the rare gas.

But for measuring very low pressures and for determining whether or not the pumping system can attain the required vacuum before filling the tube, a McLeod gauge should be used, since it is the only practical gauge for measuring vacuums of the order of 0.1 micron to 10 microns. The range over which the gauge will indicate is determined by the dimensions of the chamber D. If a large sample of gas is caught and compressed into a very fine tube, the gauge can be made to read very low pressures. In the double gauge for reading both high and low pressures, shown in Fig. 69-B, it will be seen that the micron part of the gauge has a very small

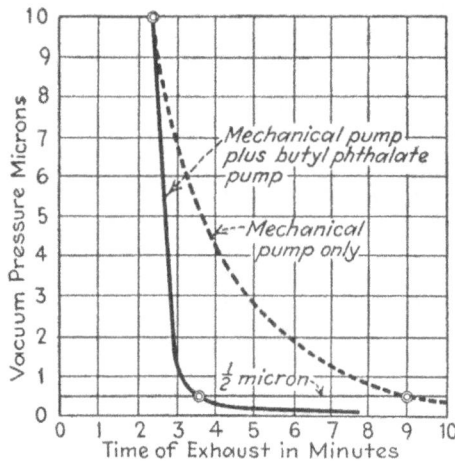

Fig. 70.—Comparison of pumping speeds, showing additional speed obtainable from butyl phthalate pump.

bore compared with that of the millimeter section of the tube.

McLeod gauges have to be very carefully made, and they must be calibrated; that is, the scale must be marked on them at the factory. It is very difficult for the glass blower to make his own McLeod gauge, and they are usually bought in complete form from a supply house.

A McLeod gauge is by no means absolutely necessary for the neon pumping plant, but for a complete system, capable of the best work, such a gauge is a very valuable instrument. It is not useful in measuring millimeter pressures, since the simpler U gauge of the mercury or butyl type is a continuously reading gauge of nearly the same accuracy. But for measuring pressures below 1 millimeter and as low as 0.1 micron, there is no alternative to the McLeod gauge except complicated ionization gauges not suited to neon work. And although it is not absolutely necessary to measure pressures as low as this in pumping and filling tubes, there is no other way of being sure that the pump will draw a satisfactory high vacuum except by measuring it.

In the newer types of McLeod gauge, the clumsy mercury flask and rubber hose have been replaced by a glass plunger, consisting of one glass tube which slides inside another tube, which contains the supply of mercury. Pressing on the sliding tube forces the mercury up into the gauge.

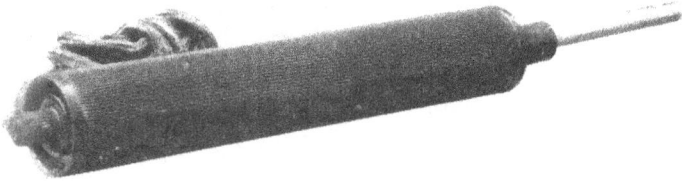

Fig. 71.—A high tension spark coil for testing the vacuum system. The knob at the end is used to control the intensity of the spark, which appears at the metal tip of the device.

Trouble Shooting. - If the system is not vacuum-tight, it may be very difficult to locate the leak. A large leak may be detected by blowing into the system through the rubber blowing hose, but leaks large enough to be detected in this manner are usually visible to the eye. The smaller leaks, which prevent the system from attaining a good vacuum but which admit air very slowly, are much harder to find. One of the most useful tools for detecting them is the *high-tension sparker,* which has been referred to previously. This device is a special transformer, mounted in a cylindrical case and equipped with a small knob (rheostat) at one end and a metallic electrode at the other. This tester, shown in Fig. 71, is attached to the 110-volt lighting circuit, and the knob at the end turned until the electrode at the other end glows with a spark discharge. The device is really a Ford spark coil built for 110-volt operation and with not quite so powerful a spark.

Use of the Spark Coil. - When the sparker is held against the glass tubing of the pumping system, the spark discharge will jump to the glass and partially cover the outer wall of the glass. A poor vacuum will be indicated by a darkpurple streak or red glow. If there is even a small pressure of gas inside the tube, a light-blue glow will form inside the tube. As the pressure decreases, this glow will become thinner and will cling closer to the inside of the glass wall. When the pressure is below 5 microns, the glow will disappear entirely.

The sparker can thus be used as a rough indicator of the pressure inside the system, and it can indicate a very high vacuum by the absence of any glow inside the tube. But it is also highly useful for locating leaks. A pale-red glow which gradually turns into a dark-purple streak inside the tube indicates a leak. The spark from the sparker will enter the tubing through the leak, and in doing so it will become very white in color as it passes through the glass. The leak will be clearly indicated by the bright spot.

The sparker must be used cautiously, since the spark is capable of puncturing the glass, especially near a bend, splice, or thin part of the glass. The tip at the end of the sparker should not be pointed, or else the spark will be very intense, and the tendency to puncture the glass will be much greater. A special type of spark coil has been developed which will not puncture the glass.

Sealing the Work to the Pumps. - When the glass blowing has been completed as outlined in the preceding chapter, the electrodes are in place, and the tubulation attached, the tube is ready for the pumps. The procedure of sealing the tube to the pumping systems is as follows: See that the stopcocks leading to the gas containers are closed and that the main stopcock leading to the pumps is closed. Then open the air-hose stopcock, admitting air to the manifold system. Break off a short section of the tubulation glass on the manifold.

Take the tube to be pumped and lay it on the pumping bench so that the tubulations on the tube and on the manifold are close together. If they do not fit neatly, the tubulation on either tube or manifold may be heated and bent until a good fit is obtained. Then splice the two pieces of tubulation glass together, using the blowtube to blow out the splice to the proper shape and thickness. When the splice is cool, blow on the air hose to test for any large leaks. If the tube and manifold appear to be airtight, close the stopcock at the air hose and open the stopcock leading to the pumps. This starts the pumping. Both mechanical and diffusion pumps should be operating at full capacity.

At the first instant the pump stopcock is opened, the sound of the mechanical pump will change to a deeper beat; but as the air is

Table XI.—Degree of Vacuum by Color, Using a Spark-coil Tester

Degree of vacuum	Color	Coil held against		Remarks
		Glass	Electrode wire	
200 millimeters....	Purple	..	✓	Color visible only at edge of electrode shell
150 millimeters....	Purple	..	✓	Faint glow in tube
50 millimeters....	Purple	..	✓	Fair glow in tube
15 millimeters....	Blue-purple	..	✓	Dark glow
4 millimeters....	Red-purple	..	✓	
2 millimeters....	Lavender	..	✓	
1 millimeter......	Light lavender	..	✓	
250 microns.......	Dark blue	✓		
100 microns.......	Light blue	✓		
10 microns.......	Very light blue	✓		
5 microns.......	Blue disappears	✓	..	No glow in tube; blue haze near glass wall.

drawn out, the deeper sound will gradually disappear. As it does so, the mercury in the U-tube gauge will begin to come to an even level. When the mercury has come down the scale tube a short distance, close the stopcock leading to the pumps. If the system is airtight, the mercury will remain motionless at the level that it showed when the stopcock was closed. If there is a leak, the mercury will begin to climb in the scale tube. The leak should be located with the sparker. When it is found, the pump stopcock should be closed, the air hose opened, and the leak repaired.

When the system is shown to be completely airtight, the tube is ready for pumping. But before pumping is started, preparations for bombardment are necessary. These preparations, together with the detailed procedure of the bombardment process, are described in the following chapter.

Pumping Hints. - Trouble with the pumping system may arise from a variety of sources. The following list of hints has been compiled to aid in running down the more common causes of trouble:

1. Keep stopcocks' surfaces greased with a thin coat of grease and keep the grease out of the stopcock bore.

2. Keep grease, oil, and water out of the manifold system.

3. Keep the oil in the mechanical pump up to the proper level.

4. Use only the purest and cleanest mercury in the diffusion pump, if of the mercury type. If of the butyl phthalate type, use pure solution.

5. Keep the pumps under vacuum at all times; that is, do not open the main-pump stopcock when the rubber-hose stopcock of the manifold is open to the air.

6. Keep the gas-container stopcocks closed at all times-, except when actually filling. Keep the piece of glass tubing between the two stopcocks evacuated. Never open the stopcock nearest the gas container until the manifold is well pumped out.

7. Never turn on the heat under the mercury diffusion pump unless the water is circulating through the cooling system and unless the mechanical pump is running.

8. Keep flame away from the stopcocks. They will crack easily.

9. Keep the sparker away from bends and splices except when testing for leaks.

10. Take down and clean the entire manifold once every four months.

11. Keep away from a mercury diffusion pump while it is operating. If it should break, the mercury vapor, which is highly poisonous, may get into the nose or mouth before it condenses.

12. Do not open air-hose stopcock until all other stopcocks are closed. Do not accidentally open the pump stopcock while blowing on the hose; the suction may injure your lungs.

13. When using a flame on the system, never heat the glass to the melting point when the system is under vacuum except when sealing off a tube. The glass will suck in if melted.

14. Keep electric motor bearings well lubricated.

15. If chain drive is used between motor and pump, keep fingers away from the chain.

Time Required for Pumping. - The average time required for pumping average lengths of tubing, with and without diffusion pumps, is shown in Table X. From this table it can be seen how valuable is a diffusion pump (even though a good mechanical pump is available) in speeding up production. The table will be found useful in figuring costs of labor arid pumping. The actual conditions in a particular plant may not require the time shown; in this case, a table similar to it should be made up, showing the actual time required which is found necessary to complete a tube of a given size. This time can then be used as the basis of cost computation.

Chapter Nine - Bombarding

The Importance of Proper Bombarding. - The commonest single cause of early failure of neon tubing is improper bombarding. Nevertheless, bombarding is not a difficult operation; if the proper equipment is available, it can be made almost foolproof. It cannot be too strongly emphasized that a little extra time and care taken in properly bombarding a tube will pay big dividends in longer tube life, even though the difference may not be visible when the tube leaves the shop.

The immediate purpose of bombarding is very simple: It is to *heat all parts* (glass and metal) of the tube to a high temperature so that the absorbed gases and other impurities will be thrown off. Trouble arises when all parts of the tube are not brought up to this required degree of heat. When parts of the tube remain below the critical temperature, the gases and impurities in those parts are not eliminated, and, in addition, these cooler parts will collect the impurities which have been thrown off from hotter parts of the tube. When the tube has been pumped, filled, and sealed, these impurities may lie dormant for many hours of operation. The tube may appear to be perfect when it is installed. But if the impurities are in the tube, they will surely be released, as soon as the operating temperature of the tube rises. The tube then loses color and must be repumped.

Bombarding Equipment. - Success in bombarding is allied to success in glass blowing. Both depend upon the skill and judgment of the operator. And, likewise, neither can be carried out by even the most skillful operator unless he has the proper equipment. Bombarding equipment is relatively simple, but it must be of sturdy design, and it must be suited to the job.

The commonest bombarding method, used almost exclusively in the neon trade, makes use of a bombarding transformer which provides a high voltage and higher current than that of the ordinary neon transformer. This transformer is connected to the tube while it is on the pumps and is used to produce a heavy glow discharge in the tube. It is thus the center of the bombarding setup. It must be supplied with current from the 110- or 220-volt power circuit; it must be supplied with an off-on switch and with some means of regulating the voltage it supplies. Lead wires for conveniently connecting the secondary terminals to the tube must be provided, and the entire installation must be insulated for the protection of the operator. In addition to electrical equipment, containers of special gases may be used.

The Bombarding Transformer. - The bombarding transformer is in many respects similar to the transformer used in lighting the tube. That is, the bombarding transformer has a definitely limited secondary current, even on short circuit. But the short-circuit current on the bombarding transformer is often ten times as high as the short-circuit current of the usual tube transformer. Tube-lighting transformers are made to deliver either 30 or 60 milliamperes on short circuit; the bombarding transformer usually delivers from 60 to 700 milliamperes on short circuit.

The open-circuit voltage of the bombarding transformer is usually from 10,000 to 15,000 volts. The transformer, while similar in operation to the tube-lighting variety, is much larger, since it must supply much more current. Furthermore, the bombarding transformer is a dangerous piece of apparatus, much more so than the tube transformer, because it has a higher secondary current. The extra

FIG. 72.—Bombarding transformer, showing magnetic switch mounted on casing, and push-button control (at end of BX cable).

current is used for heating purposes when the tube is being bombarded. To heat the tube properly it is necessary to run through it many times the normal operating current, and for this reason the bombarding transformer must be able to supply a heavy secondary current. If the operator comes into contact with the terminals of the bombarding transformer or one terminal and ground, the full 700 milliamperes of current may pass through him; *and this amount of current is definitely dangerous to life.* In order to guard against this possibility, thorough insulation of the entire secondary circuit is necessary.

The bombarding transformer is rated in kilovolt-amperes (abbreviation kva.). The rating is found by multiplying the primary voltage by the primary current and dividing by 1,000. For proper bombarding, a transformer rated at at least 5 kilovolt-amperes is required. If a smaller transformer is used, the number of feet of tubing which can be handled at the proper bombarding temperature is severely limited. Under such conditions, even average lengths of tubing will not come up to the proper temperature, and the bombarding is not 100 per cent effective.

Since bombarding transformers are relatively high-power devices, it is not economical to use them on 110-volt circuits. Usually, they are designed for 220 volts on the primary winding, since this reduces the primary current to one-half and the line losses to one-fourth. In almost every business or every industrial district, 220-volt circuits are available, and there is no reason why 220 volts should not be used for this purpose in almost every neon plant. In the rare case where only 110 volts can be obtained, the bombarding transformer can be purchased for 110-volt operation.

Bombarding Control. - In order to control the bombarding current so the proper current will flow regardless of the length of the tube and of the pressure of the air in the tube, some sort of voltage control is necessary. There are three devices which can be used for this purpose, as indicated in Fig. 73. One is the rheostat, a variable resistance which is connected in series with the primary of the transformer. This type of control wastes power but is simple and easy to use.

Another type of control is the variable reactor, or autotransformer. This device is a choke coil with an iron core. The coil is tapped at regular intervals, and these taps are connected to a set of switch points, as shown in Fig. 73-B. The lever which

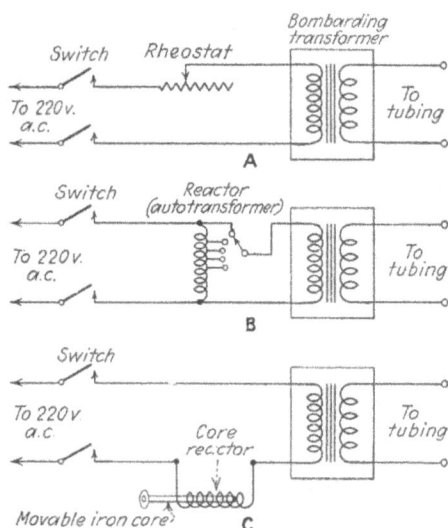

Fig. 73.—Three methods of controlling the bombarding voltage. (A) Rheostat control. (B) Autotransformer. (C) Reactor with movable iron core.

makes contact with these taps is used to control the part of the coil in shunt with the primary of the transformer and hence to control the power fed to it. This type of device uses practically no power and is used to the exclusion of the rheostat, at least in the more modern neon-plant installations. The switch

112

handle of the bombarding control should be mounted conveniently so that the operator may use it without leaving the pumps.

The third type of control is a choke coil in series with the primary of the transformer which is provided with a movable iron core. Very smooth control can be obtained by moving the iron core in and out of the choke coil (see Fig. 73-C).

Switches for Bombarding Work. - The on-off switch is connected in series with the primary circuit of the bombarding transformer. The type of switch used varies widely, from a simple exposed knife switch or a knife switch inclosed in a metal box to a completely inclosed magnetic switch, with a remote-control push button. It is recommended that the bombarding switch be carefully installed. An open knife switch is dangerous for two reasons: it exposes the 220-volt line, which can give a painful shock, and it must be pulled open to turn the transformer off. An inclosed knife switch has its metal box grounded, and is thus much safer. But the operator, leaning against an grounded box, may accidentally touch the high-voltage terminal at the same time and receive a fatal shock. A more desirable type of switch is the inclosed magnetic one shown in Fig. 72. This switch turns the transformer on when the button is pushed; when the button is released, the transformer goes off. If the operator accidentally comes in contact with the high-voltage wiring, the transformer will be turned off as soon as he releases the button. For "flashing," that is, turning the transformer on and off rapidly, this switch cannot be equaled. It is completely inclosed and safe.

Oftentimes a foot switch is used, since this leaves the hands free. But free hands can get into trouble while bombarding is going on, and it is usually wisest to use a hand switch.

High-voltage Connections. - Between the high-voltage terminals of the transformer and the electrodes of the tube there must be two high-voltage connecting wires. Since these wires are the part of the secondary circuit with which the operator may come in contact, they must be thoroughly insulated. The wire itself should be 15,000-volt sign cable, the same as is used in the sign itself. The clips which connect the cable with the lead-in wires of the tube should be of the spring-jaw type (commonly called *battery clips*), and they should be covered with rubber hoods, so that they can be attached without touching the metal of the clip. For convenience in moving these clips, two bare copper wires are often mounted on the ceiling of the shop, with copper rollers riding on them. Each connecting cable is connected to one of these rollers and thus can easily be pulled around to the proper part of the bench for connecting to the tube.

Bombarding Layout. - A convenient layout for the bombarding equipment is shown in Fig. 74. The transformer itself with the magnetic switch is mounted to one side, out of the way, and provided with heavy leads which are covered with glass tubing. These leads run directly to the ceiling where they connect with two insulated copper bars. The connecting cables hang from pulleys which ride on the bare copper.

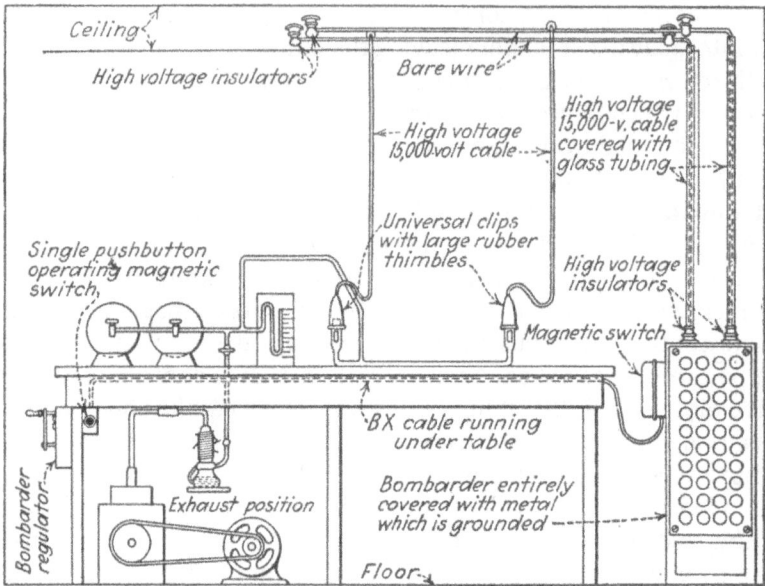

Fig. 74.—Recommended layout for bombarding equipment. The high voltage cable is strung from rollers which run on the bare wires fastened to the ceiling.

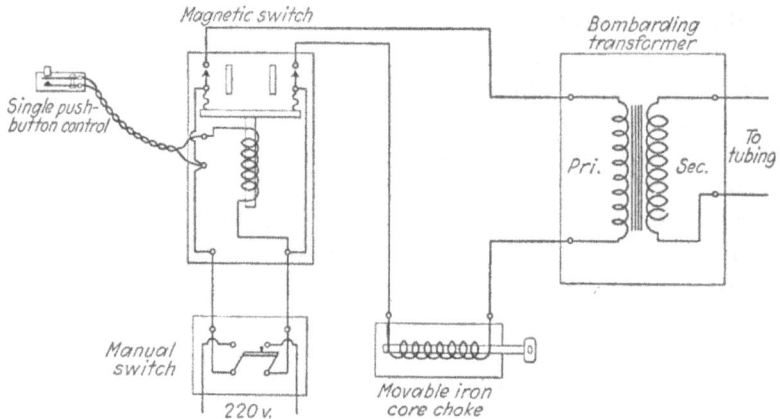

Fig. 75.—Wiring diagram for bombarding equipment, showing magnetic switch with push-button control, and movable core voltage control.

The bombarder regulator and push-button operating switch are mounted close together at the other end of the pumping bench and far enough away so that the operator is out of reach of the high voltage while the bombarder is turned on. These safety precautions should be rigidly observed. Men have been instantly killed by shocks from bombarding transformers; these tragedies need never have happened if the installations had been safeguarded and if the equipment had been used with care.

Bombarding Procedure, Simple Case. - In the simple case of direct bombarding, the following is the correct procedure: The tube is sealed to the pumps, and the entire system is tested for leaks, in the manner outlined on pages 162 163 of the previous chapter. When the system has been made vacuum-tight, air is admitted to the manifold by opening the air-hose stopcock. This stopcock is then closed.

The high-voltage wires from the bombarder are connected to the electrodes of the tube. The main-pump stopcock is then opened (with both mechanical and diffusion pumps operating). Set the bombarder control at the minimum current. As the pressure goes down, as indicated by the U gauge, the bombarder should be flashed, by giving the push button a momentary push, until a flash of light occurs in the tube. When this happens, turn off the pump stopcock immediately. Flash the bombarder again. Increase the bombarding current slowly. If the tube glows steadily, well and good; if not, open the pump stopcock for a short time until a steady glow is obtained. *Do not pump after the first steady glow is obtained.* Turn the pump stopcock off and leave it off as soon as the glow in the tube is steady.

Next take a piece of ordinary heavy white bond paper and place it against the side of the tube so that it makes good contact. Then turn the bombarder on; increase the current until the tube gets thoroughly hot; leave it on until the piece of paper has been definitely scorched to a brown burnt color. Keep an eye on the electrodes. They should get hot, so that they glow red when the bombarder has just been turned off. But they should not get so hot that they begin to sputter or glow brighter than a dull red.

When the paper has been thoroughly scorched, open the pump stopcock. The bombarder should be flashed again when the pressure has been reduced somewhat. The tube will glow as before, but the electrodes will get much hotter. Do not overheat them but get them red, while the pumping is going on. When the tube has been thoroughly heated, turn off the bombarder. When only a very pale glow occurs in the tube, do not turn on the bombarder again, except momentarily. This means that the pressure is very low, and the electrodes may be seriously injured if the high voltage is steadily connected to them under these conditions. Finally, disconnect the bombarder wires. The tube is then ready for filling.

Reasons for the Bombarding Procedure. - The above bombarding procedure for the simple case is based on sound practice. If followed, it should give complete and satisfactory bombarding under all ordinary conditions. The reasons for doing the job in the way outlined above are as follows: The tube should be bombarded at the highest possible pressure, that is, just after it begins to glow and before the pumps have had a chance to reduce the pressure below that point. The reason for bombarding at high pressure rests on the fact that the *glass* will be heated only if there is *plenty of gas* inside the tube to carry the heat to the glass wall. In the high-pressure glow discharge, most of the heat goes into the gas itself, and gas transfers it directly to the walls of the tube. If the gas were at lower pressure, the electrodes would be heated, but the glass would not become hot. To heat the glass, bombard at the highest pressure at which a steady glow will occur. To bombard the electrodes, a somewhat lower pressure should be used but not too low, or the electrodes will be heated excessively.

It may seem strange that the glass must be heated so thoroughly and carefully. But the reason is simple. The glass wall takes up about 95 per cent of all the area on which impurities may collect inside the tube; the electrodes occupy only 5 per

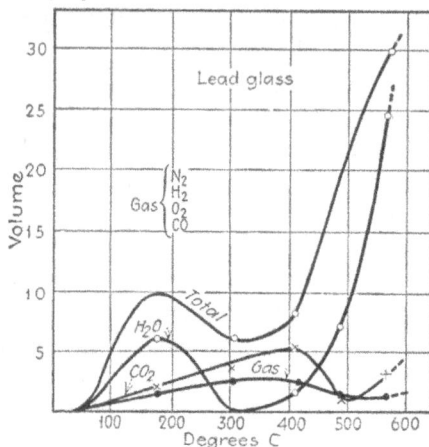

Fig. 76.—Curves showing gases released from lead glass when heated to various degrees of temperature. The following symbols are used: H_2O water vapor; CO_2 carbon dioxide; "gas" N_2 nitrogen; O_2 oxygen; CO carbon monoxide. Note that above 400°C., the released gas comes from *within* the glass structure. (*Curve due to Sherwood, Westinghouse Electric and Manufacturing Company.*)

cent. They are so small in area that they will be taken care of if the glass is brought up to the proper temperature. The scorched-paper procedure for testing the temperature of the glass is a reliable guide in bombarding, and it should always be used.

Bombarding Procedure, Special Cases. - In some special cases, the simple procedure outlined above will not do a proper job. This is particularly true when long tubes are to be bombarded or when the bombarding transformer is not quite large enough to supply the required current. In such cases, one or more auxiliary bombarding electrodes are needed. These electrodes, exactly similar to the electrodes used at each end of the tube, are

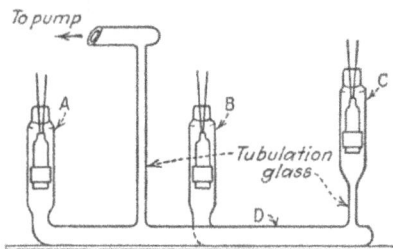

Fig. 77.—Method of connecting auxiliary electrode for bombarding long tubing units. *A* and *B* are the regular electrodes, *C* the auxiliary electrode with long tubulation, and *D* the letter itself.

sealed in to the tube with an extra-long 6-millimeter diameter tubulation. If only one extra electrode is needed, it is sealed in near the center of the tube. When bombarding is started, the middle electrode and one of the end electrodes are connected to the bombarder, and the simple procedure followed. When that end of the tube has been thoroughly cleaned out, the bombarder is connected to the middle electrode and the electrode at the other end of the tube. The procedure is repeated for that end of the tube. Finally, the extra bombarding electrode is sealed off the tube.

116

In exceptionally long tubes, two extra bombarding electrodes may be required, spaced evenly along the tube. In this case, each section of the tube is bombarded in turn until the complete tube has been covered. This procedure is makeshift, since the cool parts of the tube are always ready to absorb the gases liberated from the part of the tube being bombarded, instead of allowing them to go into the pumps where they belong. But if the bombarding transformer is not quite large enough to do the job, the extra electrode is the best remedy.

Bombarding Gas. - In other special cases, particularly with certain types of coated electrodes, a special bombarding gas is used to help bring the tube up to the proper temperature. This gas, usually pure hydrogen or a special mixture of gases, is obtained in the same sort of containers used for the rare gases, and the containers are sealed into the manifold system in exactly the same way. When the pressure within the tube has been reduced as low as the pumps will take it, and before the bombarding is started, the main stopcock to the pumps is closed, and the stopcock leading to the special bombarding gas container is opened. The gas is allowed to fill the tube to a pressure of about 20 millimeters.

When the bombarding gas has been inserted, the bombarding is started in the usual way. The gas may not have any visibly different effect from that of ordinary air, but in some types of work it is absolutely necessary for good results. It usually performs the function of combining with a chemical coating on the metal shell and reducing the salts to oxides more completely.

With certain types of electrodes it is very important not to exceed a certain current during bombardment. This is especially true for vacuum-sealed electrodes used for tubing containing helium. The

Fig. 78.—Heating electrodes by external bombarding coil, connected to high-frequency generator.

manufacturer's instructions which accompany such electrodes give the directions for properly bombarding these units.

External Bombarding. - The method of bombarding by heavy current-glow discharge, described above, is in almost universal use in neon work. But it is not the only method. For some special work, the process known as external bombarding is used. This type of bombarding is suited to heating the electrodes only; when this method is used, the glass tubing must be baked in an oven. The metal in the electrodes is heated by a process of *electrical induction,* similar to that used in the high-frequency electric furnace. A heavy coil of copper tubing is connected to a generator of very high-frequency electric current, similar to that produced in radio transmission. The copper coil is held over the outside of the glass so that it surrounds the electrode. When the current is turned on, a high-frequency current is induced in the electrode metal. As a result, the metal of the electrode becomes very hot; in fact, a white heat can be produced if the current is kept on too long.

Special Heating for Mercury Tubes. - When mercury tubes are being pumped, the mercury which is inserted in the tube must be given a heat treatment. The usual bombardment discharge will not heat the mercury, especially since the mercury usually is not inserted in the tube proper until the pumping and filling are complete. The best practice in filling mercury tubes with the required mercury is explained next chapter. The mercury container (shown in Fig. 82) is not heated by the bombardment and must be heated in some separate manner. The simplest way of doing this is to heat the mercury pool with a soft flame while the pumps are on, after the rest of the tube has been bombarded. This will free the moisture and other impurities in the mercury so that they can be drawn off by the pumps. Care should be taken, of course, to avoid heating too suddenly the glass bulb containing the mercury or heating it to too high a temperature. When heated, the mercury should boil, but the boiling action should not be allowed to become very violent.

Bombarding Tests. - There are a few tests which may be applied to the tube before it is filled to determine whether the bombarding is sufficient. They all involve measuring the pressure in the tube after it has been thoroughly pumped down. The McLeod gauge may be used to measure the vacuum, if one is attached to the system. If not, the spark-test coil may be used. If the pressure has not been reduced to a low point, it may be that there is a slight leak or that the manifold is dirty or that the pumps are not capable of producing the required vacuum. But if the pumps are in good working order, high pressure will usually indicate improper or incomplete bombarding. Filling the tube with air and beginning again from the beginning is often the only remedy. With a suitable pumping system in good working order, and if the bombarding has been carried through carefully, only one bombardment will be found necessary.

The degree of heat necessary to liberate the absorbed impurities in the tube and electrodes depends upon the type of material and the kinds of impurities to which it has been exposed. The glass wall contains two different kinds of impurities, the surface particles and the absorbed gases (see Fig. 76). The surface particles can either be converted into gas or entirely degasified by the heat of ordinary bombardment. The gas which is within the structure of the glass itself requires very high heat for elimination. Since these gases do not affect the operation of the tube, normally, they are not removed.

Caution Required. - If the bombarder is used carelessly, it will cause a great many tubes to be ruined. This is particularly true when the tube is being bombarded at low pressure, after the glass walls and electrodes have been thoroughly heated. If the bombarder is kept on too long when the pressure is low, the electrodes may become very hot and may even melt. More often, however, the discharge reaches the lead-in wires behind the electrode and melts them or else heats them to such a degree that the seal of the tube is cracked. Also, in many cases, the electrode glass jacket may be darkened by a metal deposit coming off the shell or lead-in wire. The bombarder should be kept on only for a second or two at a time except when the high pressure is still in the tube. Good practice in bombarding also requires safety precautions for the operator. A heavy rubber mat placed on the floor in front of the bombarding equipment will reduce the hazard of accidental shocks.

Chapter Ten - Filling, Testing, Aging

Filling the Tube. - When the tube has been bombarded and the pressure brought down to 5 microns or below, it is ready for filling with the rare gas. Some mention has already been made of the flasks which contain this gas, of the way in which they are attached to the manifold system, and of the way in which the gas is allowed to flow into the tube until the proper filling pressure is reached. This procedure is simple enough, since it involves only opening one or two stopcocks and keeping an eye on the pressure gauge at the same time. But if the tube is to be bright and of the proper color, great care must be taken to protect the gas inside the flask, so that it does not become mixed with air or other gases. If the stopcocks are opened at the wrong time, for example when the manifold

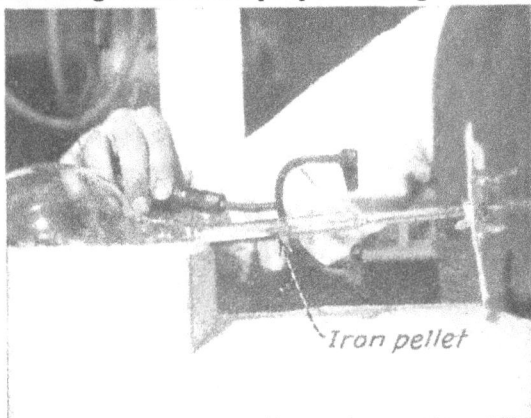

Fig. 79.—Sealing a stopcock to the rare-gas container. The asbestos shield is used to protect the stopcock from the heat of the flame. Care should be taken to prevent iron pellet (see arrow) from breaking the seal of the container.

is not pumped out, the gas from the manifold may flow into the flask and mix with the rare gas. When this has happened, the gas in the container is ruined, and there is no way of repairing the damage except by installing a new flask or by the use of a potassium purifier. It is well to understand the containers, therefore, before using them to fill a tube.

Rare-gas Containers. - The gas container shown in Fig. 79 contains one liter of rare gas at atmospheric pressure (760 millimeters). The container has two seals: one on the side of the round globe and one in the "neck." The seal in the neck is broken after the container is sealed to the system and after all air and impure gas have been pumped away.

When the bottle is first bought from the manufacturer, it is common practice to test it to see whether it is full and whether the gas is pure. Although every precaution is taken by the manufacturer to insure a full container of pure gas, the possibility of leaks' developing in the flask during the shipment makes it advisable to test the container before sealing it into the manifold.

For years it has been standard practice to test gas bottles with the spark-coil tester. When the sparker is held against a neon container, the bottle should show well-defined streaks, or streamers, of reddish color. If these streaks appear, the bottle is accepted as in good condition; if not, it is rejected. Recently, however, it has been shown that this test is by no means a sure indication of the purity or quantity of the contents. In one test, two separate liters of neon were checked by the spark-coil method. The first bottle showed streamers of red; the second, merely a red glow. When both bottles were opened and checked in a laboratory, it was found that they contained identical amounts of pure gas. One of the reasons why the spark-coil test is not always accurate is the fact that the sparker has an adjusting rheostat which permits a different intensity and frequency of spark. Even small changes in intensity or frequency make a great difference in the character of the glow in the bottle.

It has been found that the spark test can be used only for testing color. A pure-red glow of any sort, whether streamers are present or not, indicates that the neon in the container is pure. If a bluish color appears, then it is certain that the neon has been contaminated with air or other impure gas. It has also been found that it is not wise to test a bottle with the sparker more than two or three times. Under the influence of the spark, gases in the glass wall of the container may become free and contaminate the neon gas. Care should also be taken to avoid puncturing the glass of the container with the spark coil.

Installing the Flasks. - The following method should be used in installing the rare-gas bottle on the pumping system:

Lay the gas container on the workbench, with the bulb in a box or other support so that the neck of the container is parallel to and several inches above the surface of the bench, as shown in Fig, 79. Procure a glass stopcock of 2- to 3-millimeter bore. Grease it according to the instructions given previously. Prepare a small solid piece of glass or a piece of soft iron. If iron is used, heat it thoroughly in a hot flame to remove any surface impurities. Slip the glass or iron piece into the neck of the gas bottle and seal the stopcock to the bottle. Be very careful not to break the seal inside the neck. Seal the other end of the stopcock tubing to the vacuum system. Start the pumps, open the stopcock, and pump out the entire system up to the seal of the container, to as good a vacuum as the pump will draw, that is, 5 microns or less.

Close the stopcock and cut the container and stopcock off the system. Then lift the container and allow the piece of iron or glass to fall on the seal. Shake the container until the glass seal is broken. The gas will then flow out as far as the stopcock. The entire container with stopcock is then sealed back to the vacuum system and is ready for use.

If it is not convenient to remove the container from the system in order to break the seal, the iron piece may be lifted by a small hand permanent magnet or by an electric one and allowed to drop on the seal, without moving the container from its position in the system. Only an iron or steel piece can be used with the magnet for this purpose.

If two stopcocks are used on each container, as is the best practice, there should be about 4 to 6 inches of tubing between them. They may be sealed in

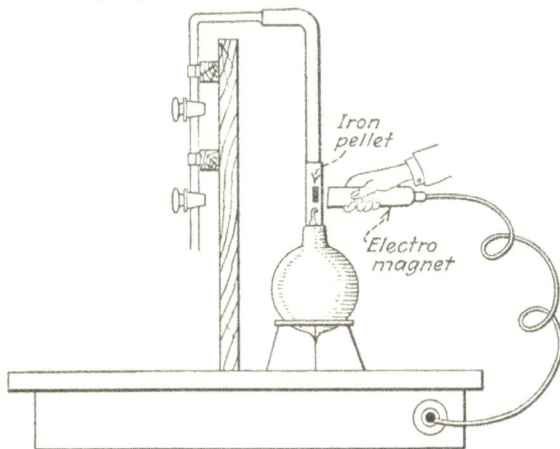

Fig. 80.—Breaking seal of rare-gas container, using electromagnet and iron pellet. The tubing between the container seal and the lower stopcock must be pumped out thoroughly before the seal is broken.

place before the pumping; or a single stopcock sealed on as outlined above, and the extra stopcock sealed on afterward.

Purity of Gas. - The important thing to remember in opening and sealing on the gas bottles is to keep the gas inside the container pure. The seal should not be opened until the space on the manifold side of the seal is under a good vacuum, and the stopcock should never be opened after the seal is broken, unless the tubing beyond the stopcock is well pumped out. If air is admitted to the container under any condition, it will mix with the gas and ruin it for neon-tube purposes. The gas will flow out if the pressure is higher inside the container than out. Air will flow in if the pressure in the container is lower than the air pressure, as it is after the container has been in use for some time.

Since the gas is precious, every precaution should be taken to protect it. The joints up to and including the stopcock should be well constructed and thoroughly annealed to prevent the possibility of cracking. If the neck of the bottle and the stopcock tubing are not of the same bore, care should be taken to flare out or pull down the glass to make a good fit before attempting the seal.

Filling the Tube. - When the bombarding procedure, as outlined in the last chapter, has been completed, the tube must be thoroughly pumped out before filling. This pumping-out process should not take more than a few minutes, but at the end of that time the pressure should be below 5 microns. If a McLeod gauge is available on the system, it should be used, at least on

121

every fifth or sixth tube pumped, actually to test the vacuum before filling. If a McLeod gauge is not used, the sparker should be used on the tube while it is being pumped. If any glow appears inside the tube, the vacuum is not low enough (see Table XI). If there is only a slight glow, close to the inside wall, the vacuum is not quite good enough, but aging the tube after filling will probably bring it up to color. If more than a very faint glow can be seen, it is almost impossible to make a good tube without considerable aging.

It is a fairly common practice to make up for a poor vacuum system by filling the tubes to a lower pressure than that recommended in Table V. This practice will produce tubes that age up quickly, but whose life is short, often only one-tenth of what it might be if good pumps and a standard gas pressure were used. This point cannot be stressed too much. *Pump the tubes down until only a barely visible glow occurs, when the sparker test is applied.* Then fill with *standard* pressures. If the pumps will not reach this degree of vacuum, or if they require too long a time to reach it, there is something wrong with them. It is false economy to use a pump in poor condition, because considerable aging is required on standard pressure tubes or, if short-lived tubes are produced, because they will run up the replacement bills.

Measuring the Pressure. - The filling pressure to be used in the tube may have been specified before the tube is put on the pumps when the original layout was drawn up,

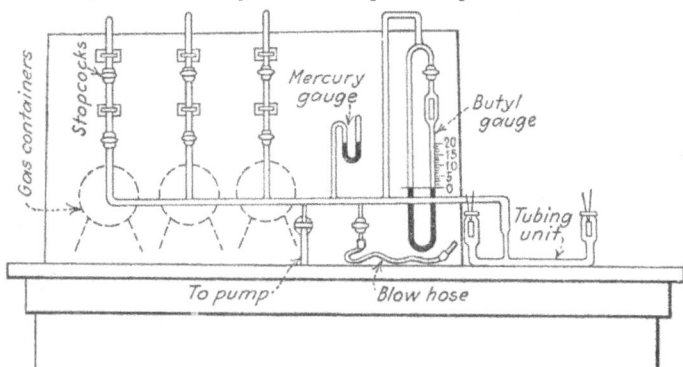

FIG. 81.—Pressure-measuring equipment, showing mercury U gauge, and butyl phthalate gauge.

according to the recommended practice. Often, however, the pressure is left to the discretion of the pump man, who must know how much pressure to use to produce the desired operating current and tube life.

When the tube has been pumped out, to the required degree of high vacuum (5 microns or less), the *main-pump stopcock is closed.* Be sure that the pump stopcock is closed before filling; otherwise the rare gas will be pumped out of the system in a very few seconds. The pressure gauge should read approximately millimeters, since the pressure is at 5 microns (only 0.005 millimeter). The stopcock of the bottle containing the gas used to fill the tube is then opened slowly. When the gas begins to flow into the tube, the pressure gauge will rise from zero, in succession, to 1, 3, 5, 8, 12 millimeters, indicating that these pressures have been reached inside the tube. When the desired pressure has been reached, the stopcock is turned off.

If too much gas has been admitted and the gauge reads high, the pump stopcock may be opened very slowly until the pump has drawn out the gas, and the pressure has been reduced to the desired value. This procedure is wasteful, of course, and should be avoided unless absolutely necessary.

The use of two stopcocks on each gas container, besides being an added precaution against leaks, helps to insure filling to the proper pressure without going over the mark. When two stopcocks are used, the lower one (nearest the manifold) is opened, so that the space between the two is thoroughly pumped. The lower stopcock is then closed, and the pumps turned off. The upper stopcock (nearest the gas bottle) is then opened, admitting a small portion of gas to the space between the two stopcocks. The upper stopcock is then closed, and the lower one opened, thus admitting the gas to the tube. The first portion of gas may not be enough; if so, a second sample is taken by the same process. After a little practice, the pump man will find that he can "ladle" gas into the tube very accurately without wasting it.

Testing the Tube on the Manifold. - In many cases, it is desirable to test the tube while it is still on the manifold. This will show whether it is perfect or will require aging. If it is not up to standard, it may be repumped and refilled while it is still on the pump. When the pump man is thoroughly familiar with his pumps, such tests are rarely necessary, but occasionally they save a lot of trouble and effort, especially if the pump man suspects that all is not well with the tube.

One method sometimes used for testing the tube while it is still on the pump is to connect the bombarder to the tube and to flash it once or twice. This sort of a test is simple to make, but it is likely to ruin the tube if the bombarder is left on too long. It should be flashed on and off quickly. The heavy current of the bombarding transformer may overheat the electrodes or lead-in wires, and the heavy current produces a glow which cannot be compared with the operating glow of the tube. The best and safest method, therefore, is to connect a standard transformer of the correct rating (either 30or 60-milliampere short-circuit current, depending on the operating current of the tube) to the tube and apply the rated voltage. If the tube is perfect, it will draw the proper current and have the proper brilliance and color. If it glows blue or blue-purple, it will require aging. This aging may actually begin on the pumps, but tubes should not be aged on the pumps, since the impure gas from the rest of the system will greatly retard the aging process. If the tube shows a purple color or does not light at all, it is defective.

Sealing Off the Tube. - If the tube tests well, or if it is judged by the pump man to be satisfactory, it is sealed off the system as follows: The tubulation glass, near the tube, is heated uniformly around the entire diameter with the hand torch for a distance of about ¼ inch from the tube. Care should be taken to heat the tube evenly and slowly until it begins to melt. When the glass is soft and is being pulled in by the vacuum, draw the tube away from the manifold, thus pulling out the melted glass and sealing the tube. When the glass has been pulled out about an inch, seal off the glass as close as possible to the tip of the tube.

It is wise to practice tipping off seals on the vacuum system before attempting to seal a completed tube, since the process is a delicate one. If the glass is heated too suddenly, it may crack. If it is heated too much in one spot, that one spot will melt first and suck in, causing a messy seal or possibly a puncture in the glass. If too heavy a seal is made, strains are set up in the excess glass causing the tip to break off afterward, often after considerable time has elapsed.

Filling Mercury Tubes. - The procedure which has been outlined in the foregoing pages for pumping, bombarding, and filling applies to all tubes, regardless of the color or gases which they contain. The mercury tube is filled with its carrier gas in the manner described, but, in addition, it has to be filled with a small amount of mercury itself. As stated earlier, the mercury tube is in a class by itself, because of the liquid mercury that it contains. Mercury is chemically active and will combine with the smallest amount of dirt or impurity present in the tube and form a thick black scum inside the tube, unless the latter is perfectly clean. In addition, the mercury vapor which provides the light of the tube is temperamental; it will not function well in cold weather. These special characteristics of the mercury tube make it necessary to follow special precautionary measures in filling the tube.

The first requirement for a good mercury tube is clean glasswork. The glass should be thoroughly cleaned before it is bent, and care should be taken to keep dirt off the electrodes and lead-in wires which will go into the tube. The second requirement is a thorough processing or bombardment. It is usually wisest to bombard mercury tubes twice, admitting air after the tube is bombarded once and repeating the process again. The third requirement is the insertion of absolutely clean mercury, and keeping the mercury pure and clean after it gets inside the tube. The fourth requirement is filling with the proper carrying gas to the proper pressure. The first, second, and fourth requirements can be met with patience and careful work. But the third cannot be met without special mercury-inserting equipment.

Methods Used for Mercury Insertion. - There are two or three methods commonly used for inserting mercury into tubes, but only one can be depended upon for successful results. Some manufacturers have a mercury reservoir on the pump system, fitted with a stopcock. When a tube is to be filled with mercury, the stopcock is opened, and a few drops allowed to flow through the manifold, by gravity, and into the tube. This system has two disadvantages: The mercury in the reservoir cannot be kept absolutely pure and clean; and, in passing through the stopcock, it is almost certain to pick up stopcock grease and dirt. Furthermore, it is difficult to measure a small amount of mercury by this method; usually the tubes so made contain an excess of mercury, which, while not actually harmful, is nevertheless wasteful.

Some tubing is made by simply pouring mercury into the tube before it is sealed to the pumps. This practice should never be followed, since it is virtually impossible to obtain a good mercury tube with so simple a procedure.

The mercury should be inserted in the tube after the pumping and bombarding are complete and after the tube has been filled with the carrying gas. Some means of heating the mercury under vacuum to free it from absorbed gas and moisture is also necessary. The system which is best suited to fill all these requirements is the side-tube method, somewhat more complicated than the methods described above but the best way of insuring a good clean tube which will last its full life without discoloration.

The Side-tube Method. - In this method, a small side tube is sealed to the back of the tube before the tube is attached to the manifold for pumping. This tube, illustrated in Fig. 82, contains a bulb for the mercury and a short length of tubulation glass which connects the bulb to the tube proper. The bulb is filled with triple-distilled mercury and sealed to the tube, bulb end down. It is important to keep the mercury out of the tube until the bombardment is over, so no mercury should be allowed to spill from the bulb into the tube. If it does so, and especially if the glass is not absolutely clean, the mercury will form little droplets on the side of the tube. Any mercury in the tube during bombardment may combine with impurities and form a black spot which it is next to impossible to remove. Also, if the mercury gets into the tubulation, it may cut off the tube from the pumps. So it is necessary to keep it in the bulb during all sealing and pumping operations.

With the side tube in place, the tube is sealed to the manifold in the usual

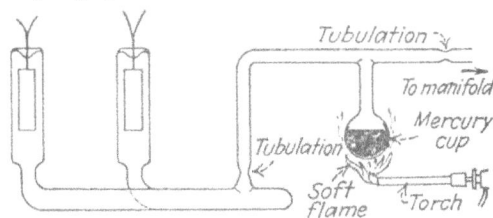

FIG. 82.—Side-arm method used for inserting mercury in blue or green tubes.

manner. The bombarding operation (double bombarding is recommended) is completed, and the pumps are used to bring the pressure down to the best vacuum obtainable. With the pumps still on, the mercury inside the side tube is then heated gently with a hand torch or other small burner, until it just begins to boil. If it contains any great amount of impurities, it may blacken the walls of the side tube during this process, but no harm is done. After the mercury has been heated, it is advisable to run the hand torch over the entire tubulation and over the glass leading to the mercury cup. This releases any moisture that may have condensed on the walls of the glass during the bombardment of the tube. When the proper vacuum has been obtained, the tube is filled with rare gas to the proper pressure (see Table V). It is then sealed off with mercury arm attached. The tube is tipped so that the mercury flows from the side-tube bulb into the tube proper. The side tube is then carefully sealed off from the tube, in the same way that the tube was sealed from the manifold. Mercury tubes made with this procedure will rarely go bad or show dirty spots or streaks.

125

Aging Methods. - A tube properly heated, with clean glass, gas, and electrodes, and properly evacuated should not require aging. But there are several reasons why aging is almost a universal process in the neon industry. One reason is that in tipping off the tabulation, impurities are released from the glass. Another is that really good attempts to speed up production have sometimes shown that a shorter time on the pumps and a longer time on the aging table are the most economical way of doing things. In the small plant, where high-speed operations are rarely possible or even desirable, this latter reason does not hold water. But for many reasons even carefully made tubing does not immediately come up to color, and it must therefore be aged.

Aging is carried out by operating the tube at rated current for a period of several minutes to

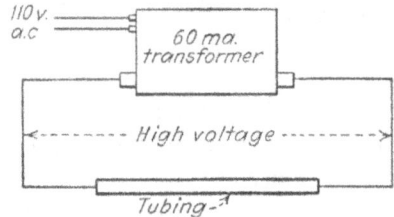

Fig. 83.—Aging with a 60-milliampere transformer.

two or three hours. During the aging process, the impure gas in the tube cleans up by chemical action, while the inert rare gas, which will not combine chemically with the electrodes, remains free in the tube. Generally speaking, this clean-up process, being a chemical reaction, can be speeded up by running the tube hot during the aging period. For this reason, often a current somewhat higher than the operating current is used for aging the tube (generally a 60-milliampere transformer is used). This extra current increases the temperature of the tube and speeds the clean-up reactions. But too high a current has the undesired effect of freeing other impurities which would otherwise remain dormant in the electrodes and the glass of the tube. So a happy medium is required to speed up the aging process without overheating the tube to the point where new impurities are liberated. For very poor pumping systems, only rated current should be used for aging, because in this case the tube is not required to stand more than the normal operating temperature; and the convenience of using a standard transformer more than makes up for the longer time required for aging.

The process of aging can be gauged roughly by watching the colors in the tube. On the first application of aging current to a neon-filled tube, the tube will probably be blue throughout the entire tube or blue only near the electrodes if it is unusually well made. This blue color almost immediately disappears, unless the tube is a mercury tube, and is followed by a bluish red. Then the ends of the tube nearest the electrodes begin to come up to color, that is, to assume the characteristic brilliant red-orange glow of neon. The red glow surrounding the electrodes then spreads toward the center of the tube, rapidly at first and then more slowly. The very center portion of the tube may not come up to color for as much as an hour or more and in some cases may not come up to color at all. Repumping is necessary if this happens.

The part of the tube nearest the electrodes cleans up first, for the obvious reason that the impure gas nearest the electrodes combines with the electrodes first. After the electrodes have received the first amount of gas, they become less active chemically, and the process proceeds more slowly thereafter.

If tubes fail to age up to color, there are several possible causes. The amount of impure gas may be so large that the electrodes cannot absorb the impurities rapidly enough, while impure gases are being released from the glass which is operating at a very hot temperature. If the vacuum system is known to be in good working order, a very slow leak at the electrodes or in the glass seals may be the cause. Or the gas in the gas bottles may have become contaminated, either with air or with other forms of impure gas. This can usually be tested by applying the sparker to the rare-gas containers. Poor electrodes may be the trouble, especially if they are not made of the very purest metal, since they then will proceed to liberate impurities at a slow but sure rate which will continually interfere with the operation of the tube. In fact, any one of the various causes which have been mentioned throughout this book as likely to cause trouble may be at the root of the difficulty.

If a tube takes very long to age (more than two hours), it is usually wisest to give it up as a bad job and begin again. The tubing may be opened, cleaned, and used again, provided that it contains no mercury and provided that the electrodes or electrode seals are not the cause of the trouble.

The Completed Tube. - When the tube has been properly aged, it is ready for installation in the sign itself. The methods used for mounting the tube in the sign depend largely on the type of sign mounting, that is, whether the sign is a box sign or a skeleton sign, a roof structure, or a border sign, a window display, or a window border, and so on. Each of these types of signs (see Chap. Five) requires a different type of mounting and a different method of handling the completed tube. Chapter Eleven deals with the mounting methods for the various types of signs.

Painting Out Crossovers. - Before the tubing sections can be said to be complete, the crossover connections between letters must be blocked out so that they do not detract from the clear-cut appearance of the sign. This blocking can be accomplished in a variety of ways. Painting is the most satisfactory method, but often a winding of tape is used to cover the crossovers.

Whatever method is used, the primary requirement, is permanence. Nothing destroys the appearance of a sign more quickly than half-covered crossovers; and if a customer complains of this fault before the guarantee is up, he has every right to service. So only the best materials should be used for this purpose.

A good tape job, covered with a waterproof varnish, has the advantage that it cannot chip off. The varnish should be of the highest grade obtainable and thoroughly waterproof. It should be applied thick enough so that the tape cannot unwind, and, as an additional precaution, a small wire may be used to bind the ends of the tape. This process, while highly permanent, is complicated compared with the use of paint and is rarely used.

There is one rule to observe if paint is used: Never use paint with a metallic base (lead paint, a copper-base paint, etc.). Metallic paints will conduct electricity. If they are used, particularly if they are used near the electrodes or a sharp bend on the tube, they are bound to cause trouble. If a voltage is built up between the paint and the near-by box housing, as often happens, a corona discharge will be set up which will attack the glass and may eventually puncture the tube. This corona discharge produces a gas known as ozone (O_3), a special

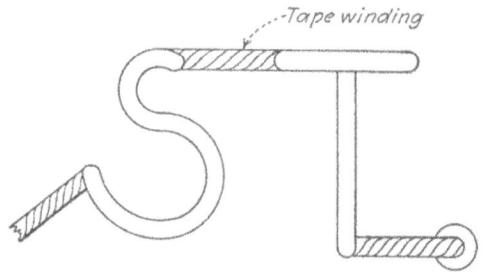

FIG. 84.—Blocking out crossovers with tape.

form of oxygen molecule which has a very destructive effect on paints. Block-out paints should have a resistance to this type of attack. Metallic paint may also cause considerable radio interference when the voltage on it discharges.

Non-metallic paints for tube painting have been specially developed and can be obtained either from a neon-supply house or from a reputable paint manufacturer. These paints should be thoroughly lightproof (opaque) when applied in a single coat. Sometimes two coats are given to insure long wear. The paints should also have a high degree of adherence to a glass surface, a quality which all non-metallic paints do not possess.

The paint generally accepted for this use is an enamel, especially made to have high insulating properties as well as to resist the weather, heat, light, high voltages, and corona discharge. Any large paint manufacturer, especially one catering to the sign trade, will be able to supply such a paint at a reasonable price.

In applying the paint, the glass should be made perfectly clean beforehand by rubbing with a wet cloth and drying. It may be applied with a small brush (¾-inch size) in the usual manner. The color of the paint usually used is black, but often colored paints are used, so that the crossovers will not stand out sharply against the sign background when the sign is not turned on.

Transporting Completed Tubes. - It often happens that the neon manufacturer simply makes the completed tubes according to specification outlined by the sign contractor. Under these conditions, the neon manufacturer often finds that he must ship the neon tubes. In other cases, the sign manufacturer may find that he has an order for a sign to be installed in a town many hundreds of miles away, and in this case also he must provide for shipping the tubes. Special cartons have been developed for this purpose and can be obtained from carton manufacturers.

Chapter Eleven - Installation Equipment

Tube-Mounting Methods, Electric Wiring, Switching, Flashers, Converters, etc.

Tube-mounting Methods. - The method of mounting the complete tube unit depends upon the type of sign, that is, whether it is a box sign, a skeleton, or a border. Likewise, the types of high-voltage wiring differ according to the type of sign. For box signs, in which the high-voltage wiring is hidden, no great pains have to be taken to make the wiring neat, so long as it is mechanically strong and electrically solid. In the skeleton or border type of sign, however, each connection between tubing units is visible to the public. For this reason, the connections must not only be strong but must present a neat appearance, and they must harmonize with the tubing itself.

Box-type Mountings. - The fundamental mounting part in all box-sign work is the elevation post. This holds the tube in place, provides the pressure with which the electrode lead wire presses against the spring in the electrode housing, and in general maintains the tubing in its correct position.

The completed tube is mounted in a box-type sign as follows: It is assumed that the mounting bracket for each elevation post is correctly located and screwed down in place on the face of the sign box. The post section is then screwed into each

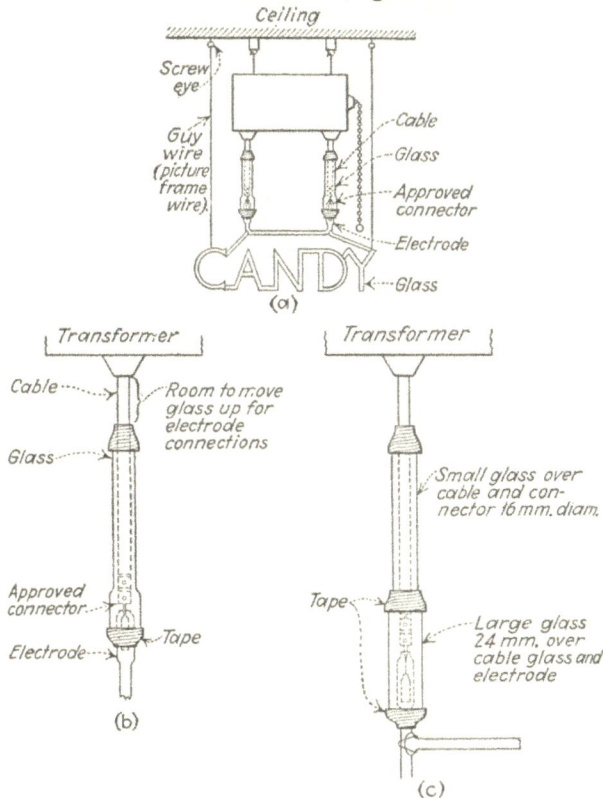

Fig. 85.—(a) Hanging transformer and skeleton sign. Note method of protecting electrodes and cable with glass tubing. (b) Detail of glass sleeve for electrode and cable protection. (c) Two pieces of tubing may be used in place of the one shown in (b).

129

bracket until the top end of each post rises to the proper level above the sign face. The tube section is then put in place, with the electrodes fitting in the two housings which have been previously inserted into the metal box. If the elevation posts are of the snap type, the tube is simply pushed into place, and the wire jaw grips the tube into its permanent position. If the elevation posts are of the more common open type, a piece of fine copper wire (not iron) is bound around the tube and under the projected shoulders of the post several times. This wire binding should not touch the sign box or any other metal in the sign.

Each tube section is fastened in this manner until the complete letter message of the sign, with borders, special designs, etc., is in place. The electric wiring inside the sign may then be put in, although most manufacturers prefer to wire the sign before mounting the tube sections, since this prevents possible injury to the tube and permits the sign box to be closed before mounting the tube sections. Wiring practice is described in detail in later paragraphs.

Skeleton-type Mountings. - The skeleton type of sign is usually self-supporting; that is, it is not solidly fastened by means of elevation posts. A small sign of this type may hang directly from its electrodes without further support, although it is always advisable to have extra wire supports from the ceiling to the glass work itself. Larger hanging skeletons may be supported on several wires, as shown in Fig. 85, with the transformer mounted above, below, or to one side of the window. In large skeleton signs, the tube is made up of several sections which must be connected together. The break in the sign between two tube sections should come between letters, of course. If it is a script-letter sign, no break at all should be visible. In this case, specially designed connections, shown in Fig. 86 (c), are used to hide the break.

The separate sections

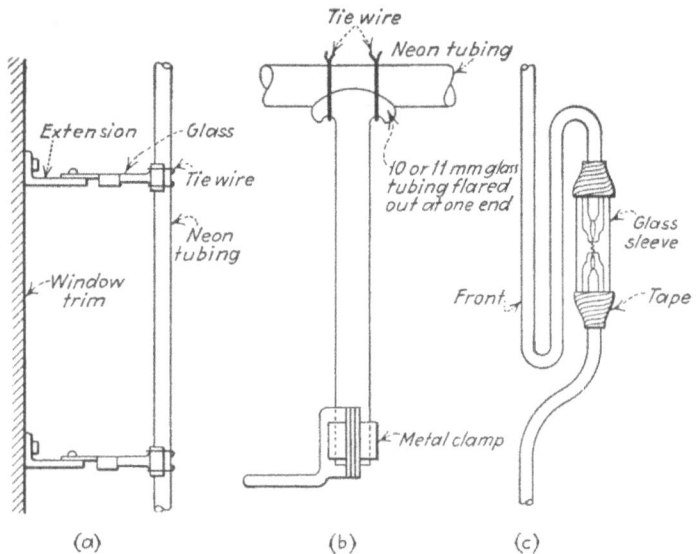

Fig. 86.—(a) Extension posts for window border signs. (b) Extension post made from glass tubing, and mounted with metal clamp. (c) Method of connecting electrode of two tubing units in a window border. No break is visible to the public.

of a skeleton sign must be strongly connected, both electrically and mechani-

130

cally. The usual practice is to slip a piece of glass tubing over the first electrode, then to connect the electrode wires together, and then slip the outer tube back so that it covers both electrodes, as shown in Fig. 85. This style of connection is neat, electrically safe, and it harmonizes with the remainder of the sign tubing. With hanging transformers, a long glass tube is used to insulate both the cable and the electrode connection. Approved metal connectors bind the cable to the electrode wires.

Types of Transformer Mountings for Skeleton Signs. - There are three ways of mounting a transformer used with a skeleton sign in an inside location:

A. A transformer can be of the hanging transformer type which is already provided with high-voltage leads, with hanging brackets, and also, in some cases, with a pull-chain switch.

B. In some cases, a hanging transformer is not desirable. A standard transformer can be used, mounted in a metal box and placed at the most convenient point either above the skeleton sign itself or to one side. In all cases, the high-voltage leads must be protected.

C. Some transformers for skeleton signs come equipped with porcelain housings. A transformer of this sort can either be placed on a ceiling with the electrodes from the skeleton sign going into the transformer or placed on the window sill with the skeleton sign coming up from it.

All skeleton signs can be classified as follows:

1. A simple skeleton sign of one color.

2. A skeleton sign which has two colors and possibly two lines.

3. A skeleton sign which has one or two colors and a border. If a skeleton sign contains more than one line and more than one color, it becomes quite complicated, and it is very necessary that it be thoroughly laid out for the glass blower so that the points of connection between the two sections are as short as possible and not visible to spoil the appearance of the whole sign.

Border Mountings. - Tubes used for building outlines or for window borders are usually supported on elevation posts (Fig. 86) since this is the most permanent and foolproof mounting. But occasionally window borders are hung from wire supports. If elevation posts are used, they are fastened either to the window trim or to a built-up metal or wood framework. Elevation posts for window border may vary in height from 2 to 6 inches. Special extensions may be used in conjunction with the standard post to give the necessary height. Hollow flares can be made from tubing in any height desired (Fig. 86 (*b*)). The tube is then bound with wire to the posts, in the same manner as in the box-type sign. The separate sections of border-type tubes must be fastened together in the same fashion as the skeleton signs, but in this case it is doubly important not to allow a break in the sign to interfere with the continuous line of the sign. The bending back of the tubing so that the electrodes are hidden behind the sign, especially if the break is offset, will prevent any break in the sign from being visible to the public (Fig. 86 (c)).

The glass sleeve which insulates and protects the connections between sign sections should be used for border signs as well as skeleton installations.

Outdoor Mountings. - Average-size signs can be transported and hung complete, with tubing attached at the factory. Long borders and glass letters that project too far above the face of the sign should be installed after the sign is hung. For large outdoor signs, such as roof structures, marquees, some large upright signs, and in general those signs that are clumsy and require a great deal of handling, the tubing is installed after the metal sign has been put in place. Elevation posts and electrode housings are used, of course, but the difficulty of transporting large signs with the tubing in them usually makes it desirable to install the tubing after the sign itself has been erected.

High-voltage Wiring, General Practice. - Regardless of whether the sign is a box-type, skeleton, or border, there are a few general rules which should be followed in all high-voltage wiring. High-voltage wiring should be kept as straight as possible. Sharp corners produce not only mechanical strain but considerable electrical strain as well. Furthermore, the high-voltage wiring should be kept at least 2½ inches away from the near-by metal box of the sign or any other metal parts.

Connections to the transformer should always be made with cable. The end electrodes of the sign are connected to cable in the manner described below. The cable is then run as directly as possible to the transformer

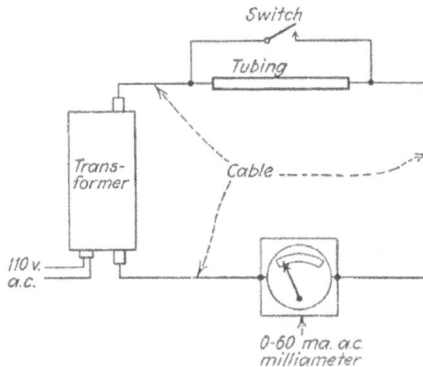

FIG. 87.—Wiring diagram for testing current output of transformer and cable. When the tubing is short-circuited by the switch, the meter should read the full rated short-circuit current of the transformer. Otherwise either transformer or cable is defective.

terminals. Terminals differ according to the make of the transformer and its size, but it will be found that a screw terminal is used in almost every case, very similar to that used on the electrode housing. A porcelain jacket or housing is often used on the transformer to cover the terminal itself; this porcelain piece will screw off, exposing the metal terminal underneath.

Cable should never be used unless it is perfect mechanically or electrically, and it should be approved by the Underwriters for the voltage used. The surface braid should not be broken, and insulation should be flexible, not stiff and brittle.

The wire inside the cable may be tested by using a piece of it in series with a milliammeter, shown in Fig. 87, and a transformer. If the wire is not broken, it will draw a short-circuit current, which will be indicated on the meter.

132

If less than the short-circuit current is indicated, the wire may be broken, and care should be taken not to use that piece of cable. When one cable crosses another, glass tubing over each cable is used to insulate the two cables from one another (16-millimeter tubing can be used).

When switches are used in the high-voltage circuit, they should always be installed in series with the complete tube and the transformer secondary.

In summary, the requirements for the high-voltage circuit are (1) a straight continuous series circuit from one transformer terminal to the other; (2) good metallic contact throughout the circuit; (3) well-insulated cable in perfect condition; (4) strong electrode connections, preferably with approved electrode housings; (5) complete protection from the weather. In wiring the sign, all wires should be kept as short and as straight as possible, and all metallic parts of the circuit should be protected from the metal parts of the sign box and its fittings.

Wiring for Box-type Signs. - If the approved electrode housings recommended by the Underwriters' Laboratories are used, they are connected as follows: The screw terminal at the rear of each housing is unscrewed to permit the cable wire to be attached. The cable itself is trimmed of its insulation with a sharp knife for about one inch from the end. The bare stranded wire is then twisted and turned into an eyelet-like form, as shown in Fig. 88, and should be soldered. The eye of this eyelet is then slipped over the screw at the rear of the housing, and the nut and washer are replaced and tightened with a wrench. It is good practice to use two flat brass washers or one flat washer and a cup washer, under

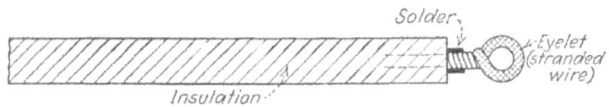

Fig. 88.—Cable end eyelet.

the fastening nut, so that it will remain permanently tight. Never use any other material but brass or copper for electrical connections, as steel or iron is not acceptable to the Underwriters. Most good porcelain housings have a flexible phosphor bronze spring to make contact with the electrode terminal. It is very important that a complete connection be made between the two. Imperfect connections produce arcing, especially in wet weather, which may cause punctured housings or electrodes or may produce high-frequency oscillations in the transformer and burn it out by puncturing the insulation.

If no electrode housings are used, usually an electrode porcelain ring is substituted for it. This ring provides no screw-terminal connection, so the lead-in wire of the electrode itself must be fastened directly to the high-voltage cable by some sort of fastener. An approved brass connector is a popular and useful device for making this connection. The electrode lead wire is placed in one end of the connector; the bare stranded wire of the cable, in the other. The two screws are then tightened, and the connection is finished. The prime requisite is a good solid metallic connection between the lead wire and the cable, a connection which will not work loose and which is mechanically strong.

Skeleton and Border Connections. - When two electrodes come very close together, as they usually do in a skeleton or border sign, cable is not used to connect them, since it would mar the finished appearance of the sign. The lead wires of the two electrodes in such cases are connected by twisting them together, or they can be fastened by means of one of the screw or spring fasteners described in the last paragraph.

A length of glass tubing is usually slipped over the connection, to insulate it and to produce a finished appearance. In border signs, where a break in the sign will detract from its appearance, tubing units should be connected as shown in Fig. 86.

The Low-voltage Circuit. - The power circuit which feeds current to the primary of the transformer is usually the 110-volt lighting circuit on the premises of the sign installation. The transformer primary is wound, therefore, for that voltage. If for some reason a 220or 440-volt circuit must be used, the transformer must be specially wound to handle the increased voltage. A 110-volt transformer will be burned out very quickly if connected to a 220-volt circuit.

The low-voltage circuit of a sign consists of the connection of a piece of BX cable or standard wiring from the current supply in the building to the sign, and the installation of an on-off switch. Also, in some cases, low-voltage automatic switches, used for turning the sign on and off or for animating the sign, are connected in the low-voltage circuit.

For window signs, cable is not necessary. Ordinary lamp cord may be used for signs with transformer rated at 75 watts or less; or special cable, which is usually provided with hanging transformers, or BX, for transformers of higher rating.

For a simple type of window sign, the lamp cord is fitted with a standard plug connection, and a simple pull-chain switch is mounted directly on the side of the metal box or even in the transformer casing itself if no metal sign box is used.

For transformers of lower than 7500 volts, no mid-point ground is provided. For higher-voltage transformers (7500 up to 15,000 volts), the mid-point is grounded to the metal case of the transformer. The metal case should then be connected to the nearest ground point. For larger installations, in which a transformer of higher than 7500 volts is used or where more than one transformer is used, standard armored cable (BX) should be used.

BX cable consists of a pair of insulated copper wires, covered with a twisted metal sheath of heavy galvanized steel. In connecting the cable, the metal sheath is cut with a hack saw, so that 6 or 7 inches of the insulated wires are exposed. The ends of these wires are then cleaned of their insulation and connected to the circuit as required. Number 16 BX cable can be used to supply circuits drawing not more than 660 watts total, for a 110-volt circuit. Number 14 wire is good for 1500 watts. These wattage ratings should not be exceeded.

The connection to the lighting circuit is usually made at the fuse box or at a junction box. The methods of locating these units and of connecting the wire to them are treated in any electrician's handbook, one of which should be consulted before work is begun.

It should be remembered that in many communities no work on the electric system of a house or business establishment may be done except by a licensed electrician. If the sign electrician is not licensed, he may have the BX cable installed by a regular electrical contractor.

The connection of the BX cable, running from the building mains to the sign, is made at the outlet box in the sign. If splicing is necessary, even in lamp cord, the splice should be soldered and covered with tape, preferably with two layers, one of raw rubber and one of friction tape.

Any low-voltage flashers (automatic on-off switches) are connected in series with the low-voltage circuit, in accordance with the instructions given under low-voltage flashing.

In a great many localities, the voltage at the sign may be low. Instead of 110 to 115, it may be as low as 100 volts. Under these conditions, if the transformers have been loaded with the maximum amount of tubing, the sign may flicker. For this reason, it is always advisable to test the voltage before designing the sign, using a standard alternating-current voltmeter, of 0 to 250 volts' range. If the voltage is below standard, allowance must be made in the layout of the tubing. In addition, the leads from the 110-volt outlet to the transformer should be kept as short and should be made of as large wire as possible, so that the voltage will not drop to any great extent between outlet and sign.

Wherever possible, the sign should be connected to the electric mains through a separate circuit, not connected to any other appliance or electrical device. If other loads are -taken from the sign supply circuit, the voltage may drop to the flicker point, when the other loads are switched on.

When a metal box-type sign is installed outdoors, usually a special metal outlet box on the outside of the building is placed at a convenient point near the sign. Approved BX cable connects from this outlet box to a similar outlet box on the sign itself.

Multi-transformer Installations. - When more than one transformer is used in a sign installation, the problem of making the circuit connections becomes more complicated than in the simple case of a single transformer. The complete wiring diagram of a three-transformer sign is given in Fig. 89. It will be seen that the high-voltage circuits are kept isolated from one another, electrically at least. The primary circuits and switching devices are connected together to the power line as shown. When more than one transformer is used, each one may be turned on independently, but it usually is desired to have them all turned on at once. For this reason, a single master switch which connects all the primaries to the power line is used.

In multi-transformer installations, the power consumed by the transformers may be very large and require special power wires. To get the total wattage consumed by the entire sign, add all the wattages of the separate trans- formers. The power circuit should be able to supply this total wattage safely; if it cannot, a larger power source must be provided.

The current in the power wires is another important figure to determine when installing the sign. If the power circuit is 110 volts, divide the total volt-amperes of all the transform-ers by 110; the result will be the line current. If the circuit is 220 volts, divide by 220 to get the line current. The value of this

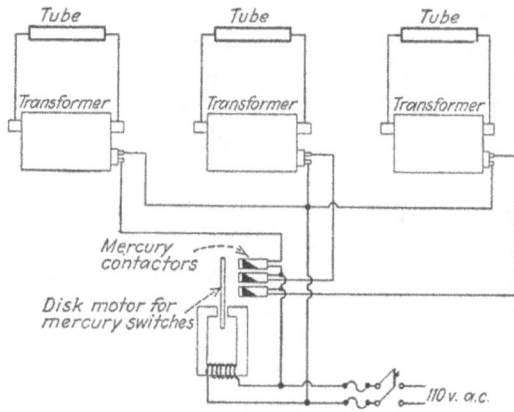

FIG. 89.—Multi-transformer connection diagram, showing three transformers with a low-voltage flashing switch for animation. A disc motor controls the mercury switches which connect each transformer in the proper sequence.

current can be used to check the size of wire used in the power circuit and the size of the fuse used. The fuse should be at least (by rating) twice the cur- rent consumed by the sign. Number 14 BX cable will carry 15 amperes; No. 16 will carry 6 amperes. In figuring the wattage loads, make sure that no other piece of electrical equipment is connected to the same fuse circuit; if it is, its wattage must be added to that consumed by the sign.

Simple Switching. - When only one switch is used on a sign, its installation presents no great difficulties. Such switches are universally installed on the primary side of the transformer. The switch is fitted with four terminals, two of which go to the power line, while the other two go to the transformer. The switch may be hand operated, or it may be automatic. Automatic time switches are installed on large signs, especially where the sign cannot be tended, as in remote locations. The time switch contains a clock, often of the self-winding variety, which trips a switch on automatically at a time set by the user, say at one hour after sunset. The switch trips the power off again at daybreak or at an earlier time, depending on how it is set. Some of these switches are almost human in their mechanical cunning; they can be devised by automatic control to set the turn-on time ahead as the days get longer during the spring months and to turn the sign on earlier as winter comes on.

A new device, recently put to use in sign switching, is the electric eye (or photoelectric tube) which can "see" daylight as distinguished from darkness and which can be used to turn on the sign when the daylight fades. This type of switch has an advantage over the time switch in that it will turn on the sign during dark cloudy days. The degree of light at which the electric eye shuts down the sign can be regulated by adjusting the circuit of the phototube. This type of device, while complicated electrically, is manufactured in complete and almost fool-proof form and can be installed easily.

Flash Switching. - The flash switch (intermittent on-off type) can be either in the low- or in the high-voltage circuit; it is a simple motor-driven switch having two terminals for the sign circuit and two for the motor drive. If it is the low-voltage type, it is installed as described above. If it is in the high-voltage circuit, it is usually mounted in the sign box, or very near the transformer if used with an open- or skeleton-type sign.

Animation-flash Switching: the Types of Circuits. – Before going into the subject of animation flashers, which is the most complicated type of electrical work with which the sign manufacturer will find himself concerned, it is useful to consider one simple element of electrical engineering, that is, the difference between a series and a parallel circuit. The series circuit, shown in Fig. 90-A, is one single electrical path, from one transformer to the other. There are no branch paths or side circuits. The parallel circuit, shown in Fig. 90-B,

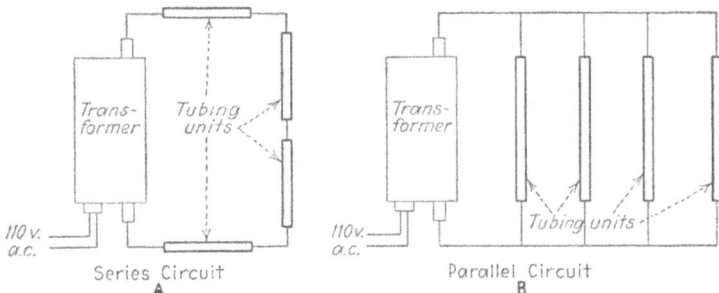

FIG. 90.—Series vs. parallel circuits. The series circuit shown in *A* should always be used for high-voltage connections. The parallel circuit shown in *B* should never be used with tubing units.

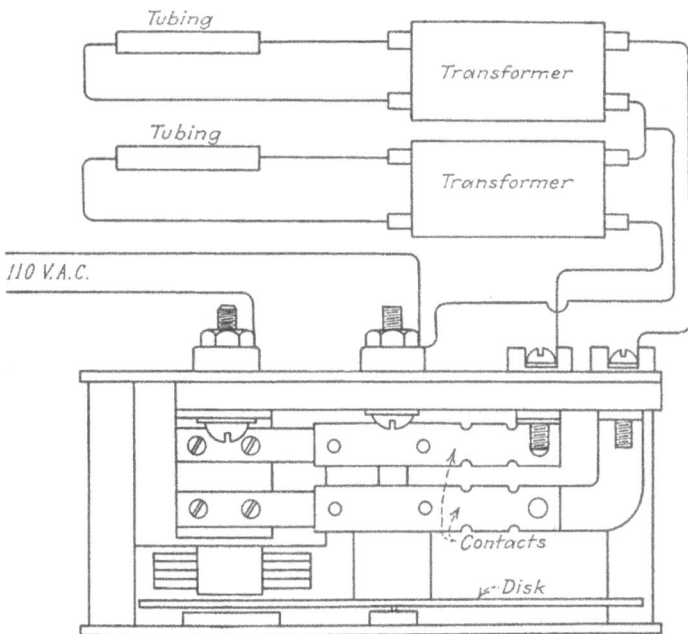

FIG. 91.—Connection diagram of a two-contact low voltage flasher. The contacts are opened and closed, in this device, by two cams which turn with the disc motor. The transformers may be turned on alternately or simultaneously, depending upon the arrangement of the cams.

consists of two or more paths from transformer terminals. The current divides into each path, the path of the lowest resistance carrying the most current.

The most important point to remember in animation switching is to connect only one single series circuit across the transformer terminals at a time. The complete high-voltage circuit of an animated sign may consist of many parallel branches, as shown in Fig. 105, but the switch is arranged so that only one of these circuits is connected at any one time. The reason why parallel circuits are never used in the high-voltage side of the transformer is simple. If there are two paths, for example, the voltage from the transformer will ionize the gas in the circuit of lower resistance and operate only that circuit. The other parallel circuit will not even light up. So the parallel circuit cannot be used.

Low-voltage Animation Flashers. - We shall treat the less common case first, that of low-voltage animation flashing. This type of flashing is suited only to multi-transformer signs, as each flasher must have its own transformer. Each transformer is switched on and off automatically and in the correct sequence by a low-voltage motor-driven switch. All the tubing on each transformer secondary gets turned on and off as a unit, and hence the number of tubing units which can be lighted independently is limited to the number of separate transformers. The

Fig. 91A.—Internal view of high-voltage rotary flasher showing disc motor, rotary switch, and contacts.

advantage of low-voltage flashing lies in the fact that the insulation of the switch can be of the low-voltage variety, which makes it simpler and cheaper. Relays may be used, also, which makes for a sturdier and more permanent installation. But for the vast majority of cases the animated sign requires a high-voltage flasher.

High-voltage Animation Flashing. - When a high-voltage animation flasher is used, one transformer may be made to serve a large number of tubing units. The diagram of a simple speller sign of eight units having one transformer and an eight-point flasher is shown in Fig. 105 (b). The flasher motor is supplied with 110-volt alternating current, through the motor supply leads. The switch contacts themselves are connected as shown. The center contact goes to one transformer terminal, while each of the outside contacts is connected to one electrode of one letter. The other electrodes are all connected together to the other transformer terminal. This type of connection lights one letter of the sign at a time.

The more usual type of speller sign is connected as shown in Fig. 105 (*d*), page 262. The letters light in succession, but each letter, once lighted, remains lighted until the completed sign is lighted. Then the entire sign flashes off, and the spelling begins anew.

In ordering animation flashers, the number of letters to be controlled and the sequence of effects required are usually sent to the switch manufacturer, who ships the required switch complete and ready to install, with a full connection diagram and instructions. It is usually best, especially if the switching is at all complicated, to entrust the problem to the switch manufacturer.

Spark-type Animators. - There is a type of animator on the market which can be used for a variety of animation effects and which has no moving parts. The animator, illustrated in Fig. 92, consists of a series of spark gaps arranged vertically one above the other. The lowest gap has the smallest separation between its contacts, and the spacing increases in each gap above. The device is fitted with ventilation holes and must be mounted vertically as shown. When the current is turned on, the spark jumps across the shortest (lowest) gap. Then, as the hot air from the spark rises to the next gap, the spark jumps to the next gap, thus connecting the second

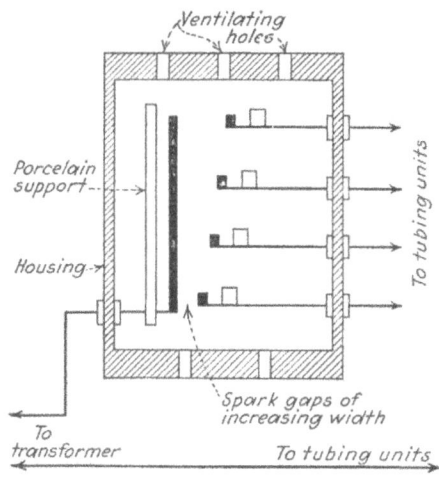

Fig. 92.—Spark-gap animator, used in high-voltage circuit for spelling and sequence effects.

section in the sign. The spark across the first gap then goes out. The spark jumps from gap to gap until it has reached the top of the chamber, where it goes out. It then starts at the lowest gap, and the process is repeated.

Hints on the Use of Flashers. - It is occasionally found that one end of a tubing unit in a sign controlled by a flasher will glow even when the sign is off. This glow is caused by the capacity of the flasher contacts, which admits a small current to the electrodes. Since the glow is usually very weak, it can be tolerated, but if it occurs, it is worthwhile examining the flasher for defects or faulty wiring.

In almost every type of high-voltage flasher, an arc is produced between the contact points as each tubing section is connected in the circuit. This arcing produces ozone, which, combining with the nitrogen and moisture in the air, is converted to nitric acid. This acid, even in dilute form, is injurious to metal parts and insulation. Protection from corrosion can be obtained by applying a coat of asphalt paint to all parts which may be affected by it.

The Cost of Electricity for Sign Operation. - It is usually the job of the sign electrician to figure the current consumption of the sign and the cost of operation. The figuring is simple enough, but a chart has been prepared which gives the cost figure without any figuring at all. This chart is explained by the caption immediately beneath it.

For those who prefer to figure the current cost accurately, the following procedure is recommended: Add the wattage rating of all transformers, switch motors, incandescent bulbs, and all other current-consuming units in the sign. Multiply this total wattage figure by the number of hours that the sign will be turned on during a month (if the sign flashes, use only the a on; ' time in calculations). This product gives the watt-hours consumed by the sign in a month. Divide this figure by 1000 to obtain the kilowatthours consumed. Multiply the number of kilowatt-hours by the cost per kilowatt-hour charged by the power company. The result is the monthly operating cost of the sign.

Example. - A three-transformer spelling sign uses one motor-driven spelling flasher. Each transformer is rated at 210 watts, primary wattage. [1] The motor-driven switch consumes 50 watts. The power company charges 7 cents per kilowatt-hour. The sign will be actually on 100 hours per month. Find the monthly operating cost.

The total wattage is $3 \times 210 = 630$ watts for the transformers

$+\ 50$ watts for the switch

_____

680 watts total

The total watt-hours per month are:

680 watts $\times$ 100 hours $= 68,000$ watt-hours

That is,

$68,000 \div 1000 = 68.0$ kilowatt-hours per month

The cost is

68 kilowatt-hours $\times$ 7 cents $= \$4.76$ total monthly operating cost

Correct Loading of Transformers. – When the transformer for the sign was selected, it was assumed that the sign would load it correctly. The transformer was chosen by reference to the data given in Table VI, according to the number of feet to be run, the diameter of the tubing, and the gas used. When the sign is actually built, it is assumed that the transformer will operate as it should without further attention. Fortunately in the great majority of cases, this procedure is a safe one to follow. But in some cases, particularly when switching systems are used, the transformer may be badly underloaded or overloaded. If it is underloaded, the transformer is not being used efficiently, and it will run hot. If, on the other hand, it is being overloaded, two difficulties may arise. The transformer will run hot; this may eventually cause trouble. But the most likely trouble is flickering. When a transformer is used to light a tube longer than it can safely handle, a radio-frequency voltage is set up in its windings. This voltage may help to cause a burnout. It may cause radio interference. It will cause the sign to flicker, especially in wet weather. All of these results are undesirable. To avoid them it is necessary to load the transformer correctly.

When spelling signs are used, the lengths of tubing connected to the transformer are not always equal; usually the length of tubing actually connected is increased as the spelling action proceeds. There is no sure way to avoid difficulties in this case, except to make sure that the transformer will safely carry the maximum amount of tubing to which it is connected at any time. Often it is difficult to predict this in advance.

All of the possibilities above may be taken as sufficient reason for testing the loading of the transformer when it is actually in the sign. Although the transformer loading test is not carried out by the majority of neon manufacturers, it is well worth while, especially for

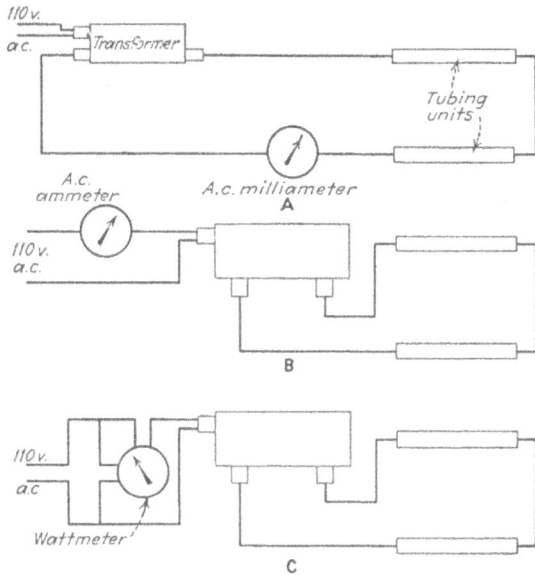

Fig. 93.—Three methods of testing the loading of transformers. (A) Milliammeter in high-voltage circuit. (B) Ammeter in primary circuit. (C) Wattmeter in primary circuit.

the manufacturer who has a large staff, where time and man power are available to do the work. There are four methods of testing the loading conditions. The first has already been described briefly; it is the use of a milliammeter in series with the secondary circuit of the transformer. The meter is used to test the current actually going through the tube. The tube footage can be adjusted until the correct operating current is drawn.

The second method makes use of an ammeter in the primary circuit of the transformer, to test whether or not it has the correct current. The correct value of this current is obtained by dividing the rated primary volt-amperes of the transformer by the primary voltage. If the ammeter indicates this current, the transformer is correctly loaded. This method is simple, provided that the alternating-current ammeter is of the correct range (0 to 5 amperes

141

is usually the most useful size), but the method is conservative and usually results in less footage being allowed than if the test were made in another manner.

The third method is the most complicated of all. It involves the making of a graph (on cross-section paper) of the wattage fed to the primary winding, against the number of feet of tubing operated on the secondary side. The point of maximum wattage will be indicated by the peak of this curve. The transformer should be operated with the number of feet of tubing to bring the primary wattage to its maximum value, less about 10 per cent. This 10 per cent leeway is allowed to take care of variations in line voltage.

The best way of measuring the wattage input to the primary is to use a wattmeter, connected as shown in Fig. 93C. This meter indicates by the position of the pointer on the scale the number of watts consumed by the transformer.

The use of this third method automatically insures that the transformer will be used at its highest efficiency and with the greatest output. If it is run at less than its rated output, it will heat up, and the result will be inefficient operation and the possibility of a burnout. If it is operated above its rated output, that is, with more feet of tubing than that which gives maximum output, the secondary current will be reduced to a low value, and the transformer will be required to stand high-voltage overloads. These may damage the insulation of the transformer, and they may also cause high-voltage breakdowns in the sign itself. In addition, as has been pointed out, if the maximum footage is exceeded, there is produced in the transformer windings and in the tubing also a high-frequency oscillation which, besides causing radio interference, is very injurious to the transformer windings and the insulation and insulating fittings of the sign itself.

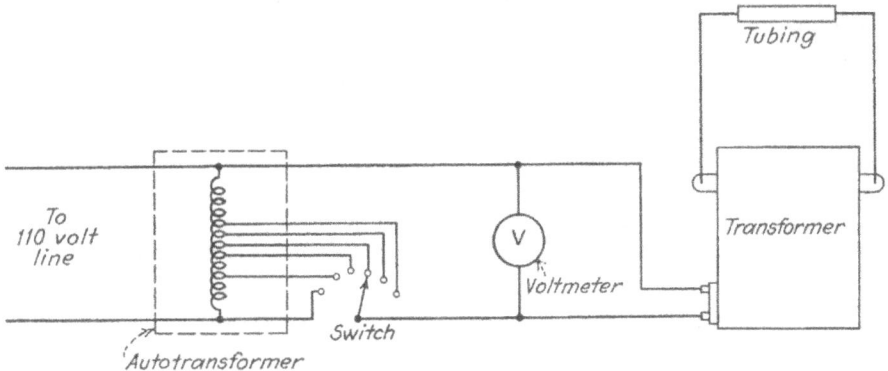

FIG. 94.—Circuit diagram used in flicker-voltage test. The switch is used to lower the voltage on the tubing until flickering commences.

The Flicker-voltage Test. - It is recommended that the completed sign be given a flicker-voltage test. This test is carried out by reducing the voltage applied to the primary of the tube-lighting transformer gradually until the tube begins to flicker and noting what this value is. The voltage reduction may be accomplished by the use of either a rheostat or a variable reactor, the rheostat being usually the most convenient method. The measurement of voltage is made with the use of an alternating-current voltmeter. The connection diagram is shown in Fig. 94. A complete flicker-test outfit can be purchased in one unit. For 110-volt operation, it should be possible to reduce the primary voltage to 90 volts or lower, before flicker occurs. If this can be done, the transformer is being operated with a sufficient margin of safety. If not, trouble may arise, and the sign may begin to flicker after it has been installed and operating for a few hundred hours.

These rules for testing the loading of transformers are given not only to check the size of the transformer when it is chosen by the use of Table VI but also to allow the proper size of transformer to be chosen when the conditions given in the table are not actually met. Differences in gas pressures used and in the diameters of tubing will produce tubing which requires transformers of larger or smaller size than that indicated by the table. In this case, the only safe way to proceed is to measure several different types of transformers on the actual completed sign and to choose for permanent installation the one which suits the job best. In the great majority of cases, the transformer can be chosen by reference to the table, but any special case requires that the above tests be made.

Installing Rotary Converters. - When a sign is to be installed on a direct-current power system, a rotary converter is used, as explained in Chap. Four. The installation of these machines is fairly simple. The machine should be mounted preferably in the basement or in some other out-of-the-way location. This will deaden the slight noise which these machines pro-duce when op-

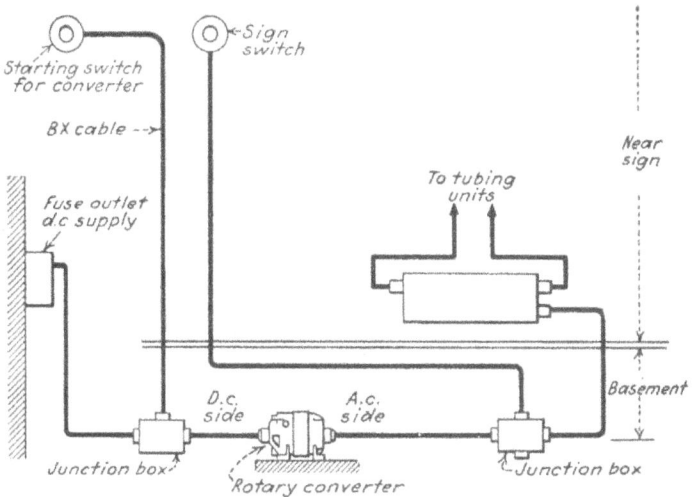

Fig. 95.—Schematic diagram for rotary converter connections. The converter must come up to running speed before sign switch is closed.

erating. The machine should be solidly mounted, but a layer of rubber under the wooden or metal base may be used to lessen the noise from vibration.

Two switches are usually used, one for connecting the direct current to the converter, which is always turned on first, and one for connecting the transformer and sign to the output of the converter after it has had a chance to get up to speed. If the switch to the sign is closed first, the machine will start

143

slowly, since it is then under load, and it may not come up to speed for a long time. During this time, the sign will flicker badly, and the transformer or rotary converter or both may be injured. It is possible to combine both switches into a single unit, so that the alternating-current switch cannot be closed until after the direct-current switch is on. Full operating instructions and directions for oiling the converter should be left with the owner of the sign.

Before installing, the direct-current voltage should be tested with a test lamp or voltmeter to make sure that it is the proper value for the converter. A test of the alternating-current voltage on the output side of the converter may also be made, but usually, if the direct-current voltage is correct, the speed of the converter will be correct, and this will insure the proper value of frequency and alternating-current voltage on the output side. Both the correct alternating-current voltage and correct frequency are of great importance. If the converter runs slow, even though the direct-current voltage is correct, the frequency will be below the standard 60-cycle value, and the transformer may be burned out as a result.

Rotary converters built for neon signs are rated in volt-amperes. It is very important to match the voltampere reading of the converter with the transformer or transformers used. Converters rated in watts will arc at the brushes and eventually burn out if used with a transformer which, although rated the same as or lower than the converter in watts, has a higher volt-ampere rating.

In all cases, the cost of rotary converters increases rapidly as the rating is increased. For instance, a 250-volt-ampere converter may cost $35. A 500-volt-ampere one may cost $50. A 1000-volt-ampere converter may cost over $100.

To get the maximum economy from converters, it is advisable to use transformers of the high power-factor type, that is, of 90 per cent power factor. In this case, a 15,000-volt, 30-milliampere transformer which has a 450-volt-ampere rating at 50 per cent power factor will have only a 250-volt-ampere rating at 90 per cent power factor. With a standard transformer, a 500-volt-ampere converter must be used; with a high power-factor transformer, a 250-volt-ampere converter can be used. The difference in cost between standard and high power-factor transformer is only three or four dollars, whereas the difference between the 250- and 500-volt-ampere converters is $15.

Protection of Exposed High-voltage Cable. - When cables must be run in the open, as in the case of almost every large skeleton sign and border sign, great care must be taken to keep them from coming too close to one another. Even though their insulation is designed to withstand the full high voltage, if the cables from opposite sides of the transformer come into contact, the strain on the insulation may be severe enough to cause eventual breakdown. In this case, each cable should be protected by a glass tube, through which it is run throughout its entire exposed length. This glass tube, besides being a well worth-while safety precaution, usually adds to the appearance of the sign.

Installation of Transformers. - Many transformer failures are caused by water's getting into the transformer casing and coming in contact with the windings. For this reason, the sign box should be constructed carefully, with tight seams and with as close a fit as possible around the electrode housings. Drain holes should be drilled in the base of the sign so that any water which collects may drain off before it can get at the transformer casing. The transformer, if mounted on the bottom of the box, should rest on channel-iron runners, so that drainage water will run under it. All transformers which are installed in outdoor locations without metal boxes should be of the weatherproof type.

One cause of trouble which is not generally understood is the failure of transformers because of metallic shielding on the high-voltage cable. High-voltage cable should never be used with a metallic sheath, for neon work. The metal casing builds up an electrostatic capacity to ground, which partly neutralizes the high flux leakage of the transformer. As a result, the terminal voltage of the transformer, even under load, rises to values much higher than normal. Burnouts, either of the transformer windings or of the tubing, are usually the result. Hence, although metallic sheathing may be attractive from a mechanical point of view, it should never be used on any high-voltage connection.

Prevention of Radio Interference. - Properly installed neon signs should not cause radio interference. When they do, some abnormal condition is present, and it should be investigated. Among the common causes of radio interference are flickering tubing; overloaded transformers; static or corona discharges between sections of the same tube, especially in double-backs and crossovers; corona discharge between tubing and ground (case or frame of sign); defective electrodes; leaky bushing or electrode receptacles; loose contacts; ungrounded transformer case; the use of conducting paint for blocking out sec-

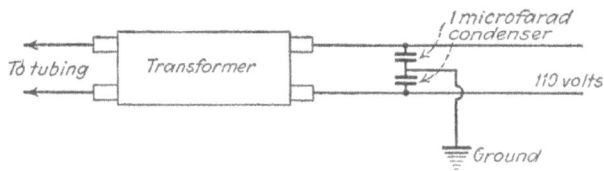

FIG. 96.—Diagram of condenser connections used for correction of radio interference.

tions of the tube; dirt or moisture on insulators or mounting parts; and sparking in flashers of various types.

When radio interference is caused by the sign, each one of the above possibilities should be investigated. If any one or more is found to be at fault, the remedy can be applied at once, in most cases. Occasionally an entire tubing section must be redesigned, so that long lengths of tubing are not doubled back upon each other or so that they do not come too close to the sign box.

The transformer itself is seldom the cause of the trouble, but if the insulation is beginning to go, it may cause bad interference. Load and flicker tests should be applied in obstinate cases of radio interference. If all else fails, the use of a double condenser across the input to the primary windings, with the

145

mid-point grounded, will usually lessen the trouble. Do not use metallic shielded high voltage cable (this remedy appeals to radio service men who may be called in by the sign owner, but it usually aggravates the trouble by raising the output voltage of the transformer). The primary wiring should be shielded, however, and the metal sheath should be carefully grounded.

[1] Be sure not to use the "K.V.A." rating instead of the wattage rating.

Chapter Twelve - Installation and Maintenance

Sign Installation. - The problem of installing the completed sign on the premises of the owner can rarely be solved by following a set of rules. Each sign and each location offers its own particular difficulties. In view of the wide variety of conditions which confront the sign installer, only a few general rules can be given to govern installation practice.

Box-sign Hanging Methods. - The installation of box signs involves sign-hanging practice, and as such it is a highly developed business. Many neon manufacturers have found it wise to procure the services of professional sign hangers who are thoroughly familiar with the work, particularly if the sign is large and heavy. Sign hanging is a dangerous business unless carefully done, and in busy districts the hazard to the public is considerable unless the work is done by competent men.

Types of Supports. *Swing Signs.* - The swing sign is supported at the top by brackets, which are built into the sign itself. From the forward edge of the hanging pole, guy stays are strung to the building wall. These stays run upward above the sign to the wall and to either side. They may be made of wire, chain, or even of solid iron. The wire or chain stays are tightened in place with turn-

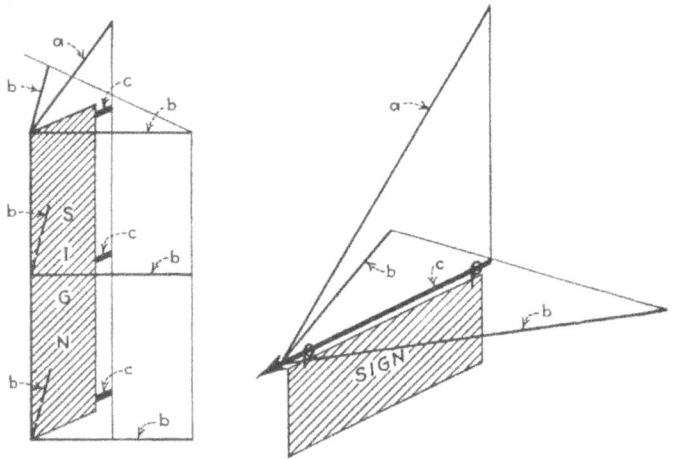

FIG. 97.—Bracing methods of upright and swing signs. (a) Top bracing. (b) Side bracing. (c) Main support.

buckles. The fastening at the wall is usually an eyebolt fastened with a lead expansion bolt. This bolt should be large enough and firmly embedded to support the sign safely, even when the sign is covered with ice or is subjected to heavy winds.

Upright Signs. - The bracing structure for upright signs is provided by channel- or angle-iron braces which fasten to the wall. The side bracing is also used to prevent swaying and should be very strong. Angle- or channel-iron cross bracing should be used within the sign itself, extending out at a 45-degree angle to be fastened to the wall.

Wall Signs. - The wall sign is fastened flat against the building wall, with expansion bolts. Stays or extra bracing are usually not required, but the fastening bolts should be strong and firmly embedded in the wall.

Roof Structures. - If the area of the sign is large, the force of even a moderate wind against it will blow a roof structure over unless it is strongly built and well fastened to the roof.

TABLE XII.—MAINTENANCE CARD

Job No. _____
Type: double-face swing sign
Guarantee: Feb. 3, 1934 to Feb. 2, 1935

Name _____
Address _____
Telephone No. _____

(north side) 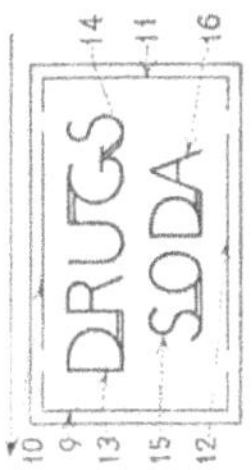   (south side)

Record of repairs

Date	Defect	Nature of defect	Cause of defect	Cost of repair		
				Labor	Material	Total
2/10/34	Fuse		Small fuse	0.50	0.05	0.55
6/7/34	No. 14*	Cracked electrode	Defective housing	1.25	0.25	1.50
10/20/34	Sign flickering	Loose connection	Poor soldering job	1.25	0.00	1.25
1/25/35	No. 11	Broken	Broken by window cleaner's ladder	1.50	0.35	1.85
			Total cost of labor and material during guarantee period			5.15

* Numbers refer to glass tubing units indicated in drawings above.

147

Projection Signs. - The projection sign (sometimes called the cantilever sign) is a modified type of swing sign with one side flush against a building and projecting out toward the street. This type of sign is usually built in combination with a wall sign, the two units blending together in a modernistic design.

Maintenance. - The service problem in the neon industry has never received the attention that it deserves. It has been assumed that the major problem was that of producing and installing the signs and that if they were properly made, the maintenance problem would take care of itself. But, sad to say, the maintenance problem rarely takes care of itself. If the reputation of the sign manufacturer with his customers is to be preserved, proper reserves both of equipment and of capital must be made to take care of servicing signs. Accidents will happen to even perfectly made signs. When they do, the manufacturer must be ready to repair the damage quickly and well. Service is paramount, since a defective sign has lost its advertising value.

The problem of maintenance can be attacked from two points of view. The first is to produce and install the signs with an eye to the future so that the calls for service will be reduced. The second is to familiarize oneself thoroughly with the major causes of failure so that when a sign breaks down the nature of the trouble can be analyzed at once and corrected with the least delay and expense. The sign manufacturer should keep both of these points of view in mind.

Causes of Service Calls. Tubing Failures. - There are two places where a sign may fail: in the transformer and its connections, including switches and fuses; and in the tube itself. Tubing failures may arise from breakage of the glass; from sputtering and cleanup of gas due to improper gas pressure; from leaks either at the electrode seals or elsewhere in the tube; from broken connections between tubes; and occasionally from blackening of the glass (especially in mercury tubes). All of these causes of trouble can be found by simple inspection or else by the application of a spark tester. They can be prevented to a large extent by the proper manufacture and installation of the tubing. Throughout the previous chapters, precautions against failure have been emphasized, and if the directions given are followed, tubing should stand up for its full life.

At the end of the useful life of the tubing, it must of course be replaced. But before the tubing unit is taken out and repumped, the precaution should be taken to determine whether the tube is really going bad or if poor connections or other electrical causes are at the root of the trouble. It is wise to take a spark coil to the job, fitted with a long extension cord. This sparker can be used to test the tubing while it is still in the sign. If the tubing unit shows no sign of life when the sparker is applied to it, then that unit is defective and must be replaced. If it glows red and shows other signs of being normal, then the trouble must be sought elsewhere. A defective tube may be caused by poor electrical connections at the electrodes. This cause of tube failure should be rectified before the repaired tubing is replaced.

Electrical Failures. - The second class of trouble is the transformer and its connections. In a recent survey of sign transformer troubles, made by one of the largest manufacturers of this type of equipment, it was found that 95 per cent of all the trouble with transformers could be traced to four causes: insufficient protection from the weather; transformer operated at incorrect frequency, as when operated from a converter running below normal speed or with excessive output voltage; improper transformer loading; and operation on too high a primary voltage.

To determine whether or not the transformer is in working order, an alternating-current milliammeter (see Fig. 87) may be connected directly across the secondary terminals. If the rated short-circuit current flows, the transformer is in good condition. If no current or less than the rated short-circuit one flows, the transformer is out of order and should be replaced.

Occasionally when a high-voltage switch is used in series with the tube, this device will cause considerable trouble. The trouble can be found almost immediately by short-circuiting the switch terminals. If the tube glows to the proper brilliance, the switch is at fault, and it should be repaired or replaced. Because of the many different varieties of switch which are used, it is difficult to give any general rules for testing them, other than the above. When the cause of the sign failure has been determined, the defective part must be removed, repaired if possible, and replaced.

Repairs and Replacements. Tubing. - When tubing is at fault, the tubing section which has failed must be removed, as a whole, and replaced. But if the sign has already reached its life limit, and especially if the still-operating tube sections show signs of age, it is often wise to take down the entire tubing and replace it. Tubing sections must eventually fail, and it will often save money to replace them all at once rather than one at a time.

The tubing section can be removed from a box-type sign when electrode housings are used, simply by removing the tie-wires which fasten it to the elevation posts and then pulling the tube out of the housing. If electrode rings are used, the connector between lead wires and the cable section must be disconnected.

Once removed, the tubing section is taken to the shop. If the glass is not broken, the tubing may be used again. In some cases, it may be advisable to replace the tubing, but usually the old tubing is still suitable for use. The first step is to remove the electrodes, which should not be used again.

Removing the electrodes is accomplished simply by breaking the glass with a file cut in the usual manner, at the joint between the electrode jacket and the tube proper. It should be remembered when making this break that the new electrode, when spliced to the tube, must fit accurately in the position occupied by the original electrode before removal and also have the same height. This necessitates care in leaving not too little and not too much glass on the tube where the splice will be made.

The outside of the glass should be wiped clean to prevent the possibility of dirt's collecting inside the tube. The tube, freed of electrodes, should next be

149

cleaned thoroughly. If it is not of the mercury variety, such cleaning may not be necessary, but it is nevertheless a wise precaution. If the tube has contained mercury, cleaning will almost always be necessary. Glass can be cleaned with plain water; but if it has been blackened from mercury stains, this simple treatment will not suffice. The hydrochloric acid solution given earlier should be used. If this method will not clean the glass, the job should be given up as hopeless, and a new tube bent.

Resealing Electrodes. - New electrodes, of the same type used in the sign originally, should be used. One electrode is spliced on in the usual manner. In this work, the original layout sheet, used for the glasswork when the sign was first made, should be used to insure the correct position of the electrode with respect to the rest of the tube. When the first electrode is on, a new tubulation glass must be spliced on in the usual manner, and finally the other electrode is sealed on.

Repumping, Bombarding, etc. - From this point on, the pumping, bombarding, and filling of the tube are exactly the same as they were when the tube was first made. Although the job is a repair, it should be done with the same thoroughness and care used on the original tube. If the tube has failed before its expected life length was reached, special precautions should be taken to see that it is repumped carefully, since the original procedure may have been faulty. To prevent a recurrence of the failure, extra precautions should be taken to make the tube in the approved manner. Often larger electrodes should be used.

Tube Breakage. - Tubes often crack under the influence of the high heat which follows gas cleanup. Also, they may break because of accidents to the tube, while it is in the sign. In these cases, it may be necessary to bend the entire tube new from fresh glass, although, when the glass is broken between letters or near the electrodes, it may be necessary to replace only part of the glass. In this latter case, all good parts of the tube are spliced into the new tube in the usual manner, the broken parts being replaced with *new tubing.

When new glass is used, it is necessary to match that of the original tube to the new glass. Usually this is not a difficult procedure, but occasionally the glass in the original tube itself has a somewhat different texture from the new glass to be spliced on to it. In these cases, the entire tubing section should be bent anew. Different diameters of tubing in the same tubing section are very conspicuous, although they may not be noticeable if the difference in diameter is small. The glass diameter of the tubing in the sign should always be matched as closely as possible; otherwise the tubing sections remaining in the sign will not blend well with the repaired section.

Matching glass is often one of the most difficult problems in repairing tube sections, especially if colored glass is used. The properties of the original glass may have become changed from being exposed to heat and to the weather. If the new glass does not fuse readily to it, the splice will be weak and subject to breakage.

Experience is the best guide in deciding whether to repair a tube section in whole or in part. Money can be saved by following whichever of these two possibilities consumes the lesser amount of material and labor and presents the lesser possibility of subsequent failure.

Electrical Repairs. - Occasionally the wiring of a sign goes bad, particularly if the cable has been exposed to the weather. In this case, it should be replaced throughout with new cable. But if the transformer, switch, or rotary converter is the cause of the trouble, the repair cannot be made except by the manufacturer of the equipment. In the ease of transformers, the job is usually so expensive that it is not attempted even by the manufacturer. When a transformer or other piece of electrical equipment goes bad, the only remedy is replacement with a new unit.

Tubing repairs may be made in a day, with no inconvenience to the sign owner. But a transformer or switch may take a week or more in shipment from the manufacturer. A dead sign during this long period is highly annoying to its owner. For this reason every progressive sign shop should carry on hand two or three transformers, at least one of which is capable of replacing the largest transformer installed in any of the signs manufactured by the shop. If the replacement transformer is large, it can be used as a substitute while a smaller one is being ordered from the factory. But the reverse is not true; if the replacement transformers in the shop are not large enough, they cannot give proper service when used to substitute for a burned-out transformer. In any case, the burned-out transformer should be eventually replaced by one of the same rating, provided, of course, that the original transformer was properly chosen in the first place. Often it is wise in replacing transformers to apply the loading tests and flicker-voltage tests outlined earlier, to make sure whether or not the new transformer is properly loaded.

Switching Failures. - When the switching devices of a sign fail, as can be tested by short-circuiting their terminals, there is usually no remedy but the installation of a new switch. If the switch is a simple on-off type, its replacement is simple. But if it is of a more complicated variety, such as a time switch or an animation switch, repairs are difficult and usually must be made at the factory. While the switch is being repaired, or while a new one is being ordered, everything should be done to

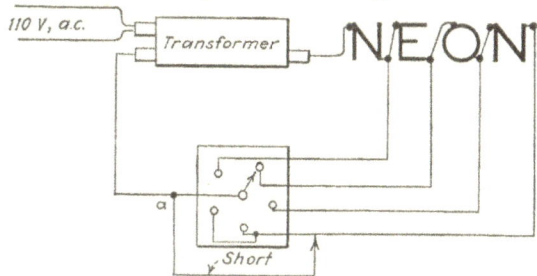

Fig. 98.—The speller switch (or other type of high-voltage switch) may be shorted out as shown if it becomes defective, until repairs can be made.

place the sign in service at least temporarily until the new part arrives.

When a time switch goes bad, a conventional hand-operated off-on switch may be substituted, with proper explanations and instructions to the sign owner so that he can arrange to have the sign turned on and off manually. If

151

an animation switch fails, the best method is to short-circuit the switch in such a way that the whole sign will be lighted continuously, as an emergency measure, until a new switch sign can be procured or the old one repaired. When this last step is taken, the owner of the sign should be consulted, and the possible increase in operating cost explained to him. Usually he will want his sign on, even though it will cost him more money for the short period.

In some animated signs, the transformers are not large enough to handle the entire sign properly when it is on continuously; in fact, if the sign is properly designed, the transformers may not be any larger than is necessary to supply the intermittent service. In this case, running the sign continuously may ruin the transformers, and it should not be run continuously until or unless larger transformers can be installed.

The Importance of Continuous Service. - In the above paragraphs, great stress has been laid upon keeping the sign operating, even by temporary measures, when interruption of service occurs. Many neon manufacturers find that they can keep their business and grow steadily without resorting to such extreme measures of service. But it is certain that customer satisfaction and the public reaction to the sign are greatly impaired when the sign fails and remains dark any appreciable length of time. Usually the customer will appreciate the effort that the neon man makes in keeping the sign lighted in the face of difficulties, and this appreciation will lead to increased business.

Neon signs are attractive only as long as they are fully lighted. A sign with several sections dead or, worse, with one or more sections weak and flickering is worse than no sign at all. The public reaction to such a display is always negative, and the sign, far from being useful for advertising purposes, actually does harm. It should be left completely dark, although a double-face sign, connected on the same transformer, may have one piece of tubing dead on one side only, while the other side still has advertising value.

In the interest of the neon business, each neon manufacturer should make this fact clear to each of his customers when he first installs his sign. He should request the sign owner to report at once any damage, flickering, or dead sections in the sign. The neon man stands to lose nothing by this procedure. Since he will have to attend to a damaged sign sooner or later, he might as well do it before any great amount of harm has been done. If the damage occurs before the guarantee is up, the service is costly to the manufacturer. But, on the other hand, if the damage occurs after the guarantee period, the result is business for him. All in all, it is a wise policy to inform the buyer that a broken sign may do harm and advise him to have any trouble corrected at once.

Maintenance Guarantees. - The average neon sign, regardless of the kind of gas used or of the complexity of the sign, is guaranteed for one year. In extreme cases of competition, this limit is sometimes extended to two and even three years. Experience has proved that many signs will give three years of life and more, but the one-year guarantee is the only sensible arrangement to make. The guarantee usually takes the form of complete service during the guarantee period, which includes replacement of all weak, flickering, or dead tubes and of any defective electrical equipment. It may also include repainting of the sign box and crossovers if either begin to shed the paint. In other words, the sign manufacturer undertakes to keep the sign operating as well and looking as well for a period of one year as when it was first installed. After the guarantee period is up, repairs are charged for on a regular basis, which includes the cost of the materials and labor and a reasonable profit.

The Service Contract. - Another form of service arrangement is finding considerable favor. It is the service contract. Under this agreement, service contracts are sold either to a man whose guarantee has expired or to a man who buys a new sign which is guaranteed for a year, at the end of which the service begins. In exchange for the service, he agrees to pay a contract price of so much per month for complete service to the sign, whether or not any repairs are necessary. This contract basis may continue for as long as the buyer wants the service, but since the sign will begin to wear out rapidly after two or three years of service, the contract rate increases as the sign grows older. In other words, in the service contract, the sign manufacturer agrees to keep the sign in constant repair at a definite sum per year, payable monthly. This amount depends on the size of the sign, the number of feet of tubing, the age of the sign, whether or not it is to be painted, and also whether or not it has a rotary converter.

The service contract is very desirable. After a sign manufacturer has been in business for some time, he may accumulate a great many service contracts which bring in a steady monthly income. By constructing the original sign so that it is sturdy and the tubing and materials of highest quality, actual servicing may be cut down to a minimum. Then the service contract is a very profitable item.

Since the contract rate calls for complete service, whether it be a simple tube section or the replacement of a whole battery of transformers, the manufacturer usually finds that it is best to install a well-made sign. Then, if it stands up, he gets the contract rate but has not any material cost or labor cost for repairs. In this way, the contract repair rate puts a premium on good workmanship in the first place. Although the man who buys the sign may feel disgruntled at paying out a monthly rate for service which his sign may not need, he has always the comfort that if trouble does arise it will be taken care of without further cost.

Estimating Repair Costs. - Before the retail price can be put on a complete sign, some estimate of the probable repair costs must be made, so that the extra amount can be added to the cost or the contract repair rate can be established. It is not the function of this book to establish business or accounting methods of figuring plant and material costs or to prescribe a fair rate of profit on the owner's investment. But the problem should be dealt with by each manufacturer.

Where competition is heavy the price situation will take care of itself; it is simply necessary to meet the quality and price of competitors and if possible to surpass their workmanship. But a fair price for sign manufacture, installation, and repair should be carefully arrived at by every progressive sign manufacturer. As an aid to this work, a complete record of equipment, materials, and labor cost should be kept, an inventory of stock should be taken regularly, and the cost attributable to each job estimated against the actual amount paid for materials, equipment, and labor. The use of the original layout in this work has been described. In Table XII is shown a "maintenance card," which is also useful in handling repair estimates.

Chapter Thirteen - Special Applications of Luminous Tubes

Special Uses of Tubes. - The sign industry is by no means the only outlet for luminous tubes. Neon, mercury, and sodium vapor lamps have been used for years for illumination purposes, as beacons, floodlights, and as lighting sources for stores and homes. Although the neon-sign manufacturer rarely has any call to make such tubes today, those companies who do manufacture them are doing a larger business every year. Eventually the sign craftsman may find a considerable market for the non-sign type of tube, and it behooves him, therefore, to investigate the uses as they exist today.

Neon for Beacon Service. - In the fields of aviation, and in the marine service, neon has recently become a highly useful source of light for beacon purposes. The high efficiency of neon in the red range and the extreme penetrating power of its red-orange color make it almost ideal for this use. Tubes for beacon service are usually of the high-current type, using hot cathodes and tube currents as high as 1000 milliamperes (1 ampere) or higher. The tubes are usually bent in horseshoe or hairpin shape, arranged in a form to concentrate the light. As many as twenty or more such tubes are often assembled in a single beacon. Lights of truly enormous power have been developed; one beacon is said to have a power of 500 million spherical candle power and to be visible in clear weather for a distance as great as 125 miles. The penetrating power of such beacons has been claimed to be far superior to that of any other kind of light. [1] The largest beacons can be seen 20 miles through fog.

A striking example of the use of these beacons was shown in tests conducted for the U. S. Air Mail Service. A large beacon mounted on top of a 150-foot tower was found to be visible to pilots separated from it by more than 50 miles of fog and mist. The British Air Ministry installed a series of straight tubes on each side of the runway of an airport landing field. A balloon containing an observer was sent aloft to 1600 feet to view the beacon under varying conditions of weather. As fog blanketed the field, the last lights to disappear were the neon tubes. Similar beacons are being used for fog service at lighthouse stations on the coast service.

Sodium Lamps for Street Lighting. - Sodium, a very active metallic element, when heated in vacuum gives off a vapor which can be used as an extremely efficient source of yellow light for general illumination purposes. This light is now used in several cities in the United States for street lighting, and it has also found use as a store-window illuminant. The sodium is so active chemically that it will attack and eat away all ordinary types of glass. For this reason, fused quartz or a special resistant glass must be used for the tube. Since the vapor is of extremely low pressure at ordinary temperatures, the tube must be thoroughly heated before conduction will begin. Neon gas is usually used to start the glow; the neon heats the tube until the sodium begins to conduct. Thereafter, the lower resistance of the sodium vapor carries most of the current, and the neon either fails to ionize or carries only a small current.

Hot cathodes of the heated filament type are used in these lamps. They have the greatest efficiency (highest output in lumens per watt) of any commercial source of light. The yellow light is claimed to be particularly suitable for highway lighting, since it reveals black objects very clearly.

Neon and Mercury Tubes for Window Lighting. - Neon and mercury tubes installed in reflecting fixtures are in wide use as floodlighting units for window displays. Coldcathode tubes can be used for this type of tube, but the high-current tubes which use hot cathodes have higher light efficiency. The combination of red and blue tubes produces a whitish light not unlike sunlight. Red tubes used for this purpose have lighting outputs of 9.5 lumens per watt or higher, while the blue tubes give 18 lumens per watt. The over-all efficiency is 10 to 11 lumens per watt.

Units of this type will operate on either alternating or direct current, since the voltage used on the tubes does not have to be stepped up by a transformer. The life of the tubes is approximately 5000 burning hours. The tube currents run as high as 3 amperes.

The tubing is hidden from the public, in this type of installation, and is installed either above or to one side of the display. Neon, mercury, and krypton blue tubes have also been used in home lighting. A hidden recess behind the ceiling molding is fitted with tubes of this type. The glow is reflected from the ceiling, down to the room.

Combination Bulb -and -tube Units. - One of the nearest approaches to artificial sunlight is obtained from combining incandescent lamps with mercury-arc tubes. The incandescent lamps are deficient in blue light, which the mercury lamp supplies; the result is a uniform balance of color very closely approximating bright sunlight. A considerable business has been built up in the merchandising of these units. They are especially useful where a good artificial light is needed, for designing, color matching, artists' studios, and so on. Stores (such as shops selling outdoor wear) and display rooms which

155

desire the effect of bright sunlight have also installed this type of illumination.

The relative amount of blue light which must be mixed with the incandescent light has been determined for the various uses to which the units are put. Usually about one-quarter of the wattage of the entire unit is consumed by the mercury bulbs, the remaining power being taken up by the Mazda lamps.

The type of mercury tube used is the Cooper-Hewitt lamp, described in a later paragraph. The tube current is approximately 3 amperes. By the use of various shapes of tubing the fixture containing the tube and bulb can be made in almost any shape, as a chandelier, hanging fixture with an opal-glass cover, or any of the conventional ceiling-, wall-, or pedestal-type fixtures. The mercury tubes have an output of approximately 14 to 15 lumens per watt, while the high-wattage Mazda lamps run from 18 to 21 lumens per watt. The mercury lamp is rated at 4000 hours burning time but will usually outlast this limit. The tube is replaceable in the fixture by the use of special clips, so that it may be replaced when it loses brilliance near the end of its useful life.

Color Matching with Carbon Dioxide Tubes. - The white light produced by the early carbon dioxide Moore tubes is a very close approximation to bright sunlight. Because of the short life of carbon dioxide tubes they have no commercial use, except for color matching, an application for which they are still occasionally used. These lamps are often fitted with the electromagnetic valve invented by Moore which admits more carbon dioxide as the gas cleans up.

The Cooper-Hewitt Lamp. - The familiar Cooper-Hewitt lamp has been used as a lighting source in factories for many years. In addition to its use as an industrial light and as a source of ultraviolet light, it has been used in blueprinting establishments.

The Cooper-Hewitt lamp is, strictly speaking, an arc lamp; that is, the electrodes of the lamp supply the major portion of the electrons which cause the discharge. In the Cooper-Hewitt lamp, the supply of electrons does not come from a heated cathode but from a "hot spot," or glowing point, in a pool of mercury at one end of the tube. When a high voltage is applied between an electrode and the mercury pool, the discharge is started, and there appears a spot of incandescent mercury in the pool, which supplies the electrons.

The hot-cathode spot depends for its action upon a continual flow of current through the tube. In the simple direct-current form of Cooper-Hewitt lamp, the arc is started with a high voltage and is maintained by the continuous flow of current. In the alternating type of lamp, however, some special means must be taken to prevent the arc from going out while the current is reversing.

The common form of alternating-current Cooper-Hewitt lamp contains three operating electrodes and a starting ring. One electrode is a pool of mercury and is called the cathode. The other two are located at the other end of the tube, and act as anodes. The mercury cathode is connected to the center

156

tap of a transformer, while the two anodes (or positive electrodes) are connected to the two ends of the transformer winding. In this way, at least one anode is always positive with respect to the mercury pool. By means of a choke coil in series with the cathode, the current is prevented from going to zero when the voltage is zero, and in this way the arc is maintained continuously. The current through the tube is pulsating direct current, of twice the frequency of the alternating voltage with which it is supplied.

The tube operates at a very much lower voltage than the required starting voltage. Automatic starting is made possible by the use of a special external electrode, or starting ring. This ring is so connected that a high starting voltage (produced in an inductance coil) can be applied between it and the mercury pool. This high voltage starts the arc, which is maintained thereafter by the two anodes in the manner described above.

To prevent excessive currents, resistance units are usually connected in series with the lamp which produce a voltage drop if the current tends to increase. The current of the mercury glow discharge, used in mercury tubes for sign purposes, rarely runs higher than 60 milliamperes. The arc discharge of the Cooper-Hewitt lamp, on the other hand, runs from 2 to 5 amperes, that is, as high as 5000 milliamperes. The voltage required for this latter case is, however, low, the cathode voltage drop being only 5.3 volts, so that the power required for the light is not excessive. The light efficiency of the glow, 14 to 15 lumens per watt, is high, as luminous tubes go.

The depreciation of mercury arc lamps during operation has been carefully investigated. It has been found that after 1000 hours of operation, the light has fallen off to about 83 per cent of its initial value; and after 4000 hours, to about 73 per cent. The tube is usually replaced when it has fallen off to 65 per cent, since then only 65 cents of every power dollar contributes to the light.

Rare-gas Arcs. - Whenever a hot cathode is used, as in neon-tube aviation beacons or in other high-current tubes, the light is produced by what is known as an arc rather than by the glow discharge produced in the sign type of tube. The spectrum colors emitted by rare gases under arc conditions are substantially the same as those produced by the glow discharge, but the color is much more intense.

Photographic Arc Lamps. - The high output of the mercury lamp in the blue-violet and ultraviolet regions makes this light ideally suited to photography. Photographic enlarging machines make use of such lamps, and they are often used to illuminate the display windows of photographers. The largest mercury arc lamps are used for blueprinting work. Typical lamps for this work are more than 60 inches long, require a tube current of 15 amperes (15,000 milliamperes), and operate on 110-volt direct-current circuits with a power consumption of 1650 watts.

The Mercury Arc as a Source of Ultraviolet Light. - For medical purposes, the mercury arc is almost universally used when the beneficial ultraviolet rays of sunlight must be produced artificially. The glass envelope of most lamps is opaque to the strong ultraviolet rays, and, consequently, if a large output is wanted in this region, the tube is usually made of fused quartz. Such lamps must be handled with great care, since the rays that they give off are very dangerous. Patients should be exposed to them only for short periods and under a physician's direction; otherwise serious burns may result.

The Neon Fuse. - A recent non-illumination application of neon tubes is their use as a high-voltage fuse. A small tube, fitted with two electrodes and filled to low pressure with neon gas, has the property of breaking down when the voltage applied between the two electrodes exceeds a certain limit. Immediately after the discharge begins and thereafter, the voltage between the two electrodes falls to a much lower value, as explained in Chap. One. If such a tube is connected across a sensitive electric meter, for example, or other device which must be protected from high voltage, the glow discharge will not occur under ordinary conditions. If a high-voltage surge does occur, the discharge immediately starts, and the voltage across the meter is reduced to a safe value. The light of the tube also acts as a telltale signal, showing that the voltage has risen above the safe value and informing the operator of the equipment that some adjustment is necessary.

Neon Used in Control Tubes. - The neon grid-glow tube is a special vacuum tube, not used for illumination purposes but capable of controlling machinery in much the same way that the common radio vacuum tube does. The glow within the tube is started by a third electrode in the tube, and the glow current is used to control some external circuit. The *thyratron,* or mercury-controlled rectifier, works on the same principle but contains mercury vapor in place of neon. These tubes are not, strictly speaking, luminous tubes, since the light is not used, but they represent an important use of gas-filled discharge devices in industry.

Mazda-type Neon Glow Lamps. - Small, low-power neon lamps capable of operation on commercial 110-volt lighting circuits without the use of a transformer are also available to the public. These lamps contain two electrodes spaced by only a few thousandths of an inch. The light results from the cathode glow, since the voltage is not high enough to produce the positive column discharge which is the basis of the light of the ordinary neon-sign tube. A high resistance is inserted in the base of these bulbs to limit the current to a few milliamperes. The light can be used as a signal light or as a night light in homes. The power consumption is usually of the order of one watt. These lamps are also used as indicators for high-frequency voltage and in measurements of radiofrequency currents.

Stroboscopic Light. - Mercury tubes, and to a lesser extent neon tubes also, are being used to provide sources of what is known as stroboscopic light. The stroboscope is a tool which permits the examination of rapidly moving rotating or reciprocating machinery. It consists of a light source which can be rapidly turned on and off at any desired rate. The number of flashes per minute is usually set at the number of revolutions per minute (or reciprocations per minute) of the machinery The machinery is viewed in semidarkness by the light of this flashing lamp. When the lamp flashes are exactly timed with the revolutions of the machinery, the revolving part of the machine appears to be standing still, and its action can be studied at will.

The action of the stroboscope can be explained as follows: The rotating part is illuminated only once during each revolution, and at the same position in each revolution, since the flashing lamp and the machinery are operating in step with each other. The eye sees the rotating part only at the instant that the light flashes, that is, when it is in a given position, say with a spoke A upward (see Fig. 99).

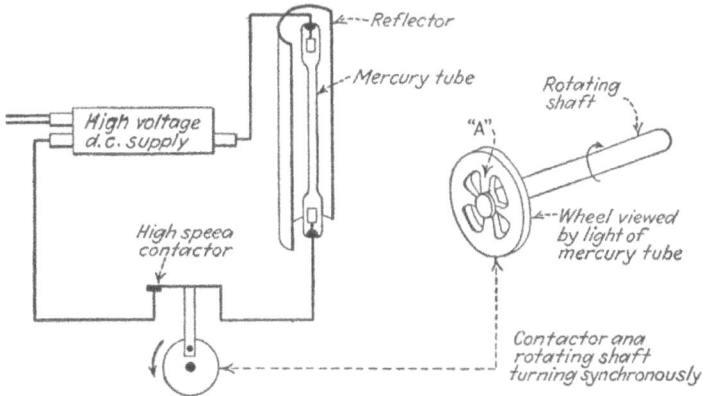

Fig. 99.—Stroboscopic light used in viewing a rapidly rotating wheel. The wheel is illuminated only when spoke "A" is upward, hence it appears to be standing still.

It sees nothing after the light goes out until spoke A is again in an upward position. The wheel appears, therefore, to be at rest, and any strains or vibrations set up by the rotation appear very clearly.

The main requirement for useful stroboscopic light is that the flash be very quick and brief, so that the wheel is illuminated only for the short instant in which spoke A is upward. Ordinary incandescent lamps cannot be flashed quickly enough for this purpose, but almost any kind of gas-discharge tube starts and stops so quickly that it is ideally suited to the purpose. Mercury has been used because large light outputs can be conveniently obtained from it. But the greater convenience of neon has led to the development of small neon tubes for this purpose. A motor-driven contactor switch is often used to flash the tubes on and off at the proper instant.

Other Uses. - Gas lamps are used in television and in the making of motion pictures. In television, the neon (recently carbon dioxide has been used because of its white color) lamps are used to produce the picture by flashing at appropriate intervals as a disk is rotated in front of the lamp. For sound motion pictures, a "crater" lamp (an argon-gas mixture is usually used for this purpose) is used to produce the sound tracks on the film. This light flashes in accordance with the sound waves and exposes the film as it passes the lamp.

[1] There has been considerable controversy over the relative usefulness of neon and incandescent lights for aviation service. Breckenridge and Nolan, experimenting for the U. S. Bureau of Standards, have proved beyond doubt that red light (either from neon or from any other source) is not more penetrating than

white light of the same candle power and in fact have proved that white light penetrates farther. In the *Bureau of Standards Journal of Research*, they say: "The addition of a red filter does not increase the range of a clear beam under any weather conditions. However, red and neon flashes...were...easier to find among [other] lights." They conclude that when a red light is to be installed, if the neon lamp can be constructed compactly, so that large reflectors are not necessary to obtain the high brilliance required, and if the accessory apparatus is not clumsy or expensive, the neon lamp "will merit consideration on the basis of efficiency alone and will stand superior to the incandescent lamp and filter." Where there are no competing lights, the unfiltered incandescent lamp has better visibility. But where competing lights are met with, the red lamp's lower visibility is compensated for by the greater ease with which it is picked up. This effectiveness of neon in competition with other lights adds greatly to its value as an advertising medium.

Chapter Fourteen - Tricks of The Trade

In the interest of brevity and clearness, the contents of the foregoing chapters have been limited to essential subjects. There are many trade practices observed by most experienced manufacturers, however, which while not essential to the manufacture of tubing are nevertheless of great value. This chapter is designed to bring together some of these tricks of the trade for the use particularly of those who have had experience in sign building.

Glass Blowing. - When working with colored glass, particularly with red or opal glass, it will be found that more heat is required to melt the colored glass than is required for clear glass. When sealing colored glass to clear glass, therefore, it will be found that the clear glass will melt first and that the colored glass will merely stick to the clear glass. Heat the colored glass to a molten state first, and then the clear glass, so that both flow together. A good joint can then be made without further trouble.

Tubulations should be made from 5- or 6-millimeter glass, and the full opening of the tube should be preserved. Any constriction in the tube will lower the pumping speed.

Asbestos layouts may be made with one of two kinds of asbestos paper. The standard 12or 14-pound stock asbestos paper is widely used. But another type, known as the 020 (or 10 Ib.) nonburn paper, has a tougher fiber and is much thinner. Although the nonburn paper is somewhat more expensive, it is often more economical in the long run, since it provides so much more surface per pound. For example, 020 nonburn paper runs 135 linear feet per 100 pounds, while the 14-pound stock runs only 85 linear feet per 100 pounds.

Pumping and Pumping Systems. - When a stopcock has only a short length of tubing attached to it, it may be very difficult to seal it to the manifold, since the heat from the torch will often crack the stopcock itself. When such is the case, a shield may be set up between the flame and the stopcock. A circular piece of asbestos paper, with a hole in the middle just large enough to slip over the tubing, may be used. This will keep much of the heat away from the stopcock. Heat conduction through the glass itself may cause trouble unless the work is carefully done.

Stopcocks may be cleaned by the use of alcohol and a clean cloth which has no lint or fuzz. Tissue paper moistened with carbon tetrachloride may be used also. Ordinary pipe cleaners are useful in

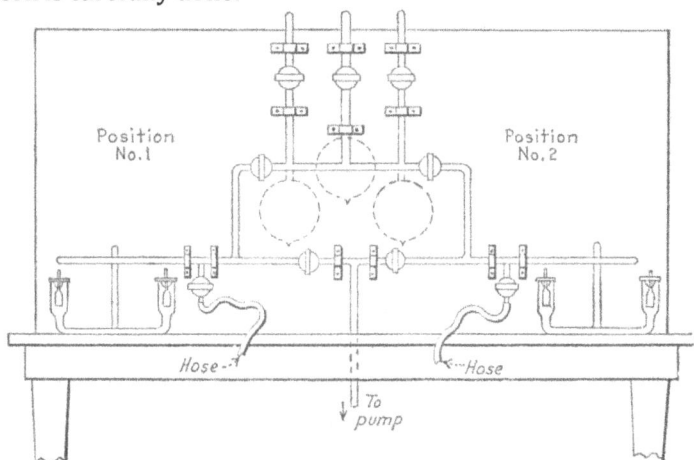

Fig. 100.—Two-position pumping system, permitting independent pumping an filling of two tubes.

cleaning out the bore. Before greasing the stopcock, be sure that no lint or dirt is left on the ground glass surface; otherwise it is almost impossible to obtain a tight joint.

When a great deal of work is to be done in a single plant, a two-position pumping system is often advisable. Such a system has one main pump and one set of rare-gas containers but duplicate stopcocks for the two positions and also stopcocks for isolating the one position completely from the other, so that work on one will not interfere with work on the other. A single operator may work both positions, attaching one tube while the other is being pumped out.

Only the lightest touch should be necessary in turning stopcocks on or off. The plug of the stopcock should not be pushed into the seat, since vacuum will tend to keep it in place.

When pyrex tubes are to be pumped on a lead-glass system, the work can be attached to the pumps by one of three ways: by the use of the graded seal; the ground-glass joint; or a high-vacuum rubber tube, which is used to connect the pyrex tube to the lead glass on the manifold directly. When the rubber tube is used, coat the two glass tubes with a thin coat of castor oil. This will seal the joint, and no leaks will occur. The rubber tubing should have a wall thickness of at least ¼ inch; otherwise the vacuum will cause it to suck in.

When a tube comes off the pumps in bad condition, the following tests should be made in order:

1. Test the vacuum of the pumps, with a McLeod gauge or spark coil.

2. Test stopcocks for leaks.

3. Test the rare-gas containers.

4. Test for leaks in the manifold.

If the system is in good order, examine the tube itself.

5. First, the electrodes.

6. Examine the tubulation tip.

7. Test for leaks in the tube itself, particularly where the letters have been spliced together.

Filling and Aging. - When a neon-gas container has been contaminated, because of a leaky stopcock or through improper use, the gas may be purified by the use of a potassium purifier, shown in Fig. 101. The rare-gas bottle is connected to the purifier, which contains two potassium electrodes. A high-voltage source (a 3000-volt sign transformer will do) is connected to the electrodes. The resulting gas discharge is accompanied by the cleanup of the impure gases, which combine with the potassium electrodes. After the rare gas is purified, it may be admitted to the tubing directly from the purifier itself, which may be permanently installed on the system.

Colored-glass tubing will often require long aging, especially when red glass is used. Very careful pumping may cut down the aging period. The impurities which make the aging necessary are probably freed from the metallic oxides used in the glass to produce the desired color.

When a neon-filled tube shows the characteristic red-orange color, it is not necessarily true that there are no harmful impurities present. A special viewing glass may be made using two thicknesses of transparent colored cellophane, one dark blue and the other dark green. These may be pasted against a cardboard to make a viewing "eyeglass." When a neon tube is viewed through this eyepiece, the impurities will show up as various shades of blue or blue-green. If the neon is pure, only the red color will be seen. For the most accurate investigation of gas purities, a pocket spectroscope should be used. These devices show the various spectrum colors to the eye and are provided with a wave-length scale so that the observed spectrum may be compared with the standard for the gas under test.

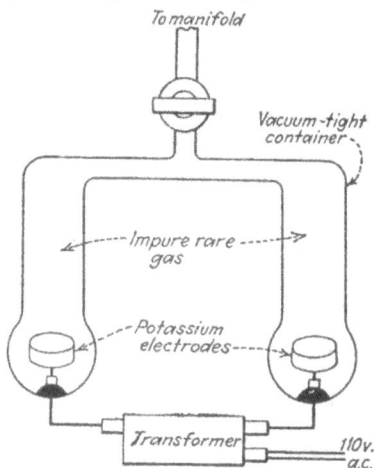

FIG. 101.—Potassium purifier for reclaiming impure rare gas.

Electrical Practice. - When luminous tubing is used in conjunction with electric light (incandescent) bulbs, on the same sign, the tubing must be placed at least 3 inches away from the nearest bulb. Otherwise, the high-voltage field surrounding the tube will cause the filament in the bulb to vibrate 120 times per second, so that the filament will break in a very short time.

162

When neon and bulbs are used in the same sign, the neon tube must be provided with a complete electrode housing, and the housing itself should be covered with a grounded metallic box. This will prevent accidental contact with the high-voltage terminals when the incandescent bulbs are being replaced. All exposed parts of the secondary, or high-voltage circuit must be covered in this way, according to the Underwriters' Specifications.

When installing signs in foreign cities, or wherever the voltage and frequency of the power system are not known, they must be determined before constructing the sign. It is best to inquire from the power company, before making the sign. When the voltage and frequency available at the point of installation have been determined, the transformer can be ordered accordingly.

It is important to check the number of feet of tubing attached to a transformer, whenever a transformer burns out. The burnout may have occurred due to overloading or underloading. If the footage is not correct, the transformer which replaces the burned-out one should be such that actual footage in the sign neither overloads nor underloads it. Reference to Table VI will indicate the correct size of transformer to use for the actual footage on the sign.

A single section of defective tubing may cause all the sections connected in series with it to flicker. The spark tester will often indicate which of the sections is defective. Otherwise each section should be short-circuited, one at a time, until the defective one is located.

Repairs and Replacements. - One of the commonest causes of cracks in or around the electrode jackets of a tube is defective porcelain housings. These porcelain units often become cracked and allow water to collect inside. In every tubing failure, therefore, it is wise to check the housings, and if they are defective they should be replaced.

Sputtering of the electrodes is indicated by a heavy metallic coating inside the glass of the electrode jacket. If sputtering occurs after the tubing has been in service only a short while, it is probable that the pressure in the tubing is too low. This is such a common fault in tubing that it cannot be too strongly emphasized. Use the recommended pressures on all tubing, or early failure will result.

Some sign manufacturers place a separate 110-volt outlet on the sign, so that when the spark coil is used for testing for defects, the power supply will be ready and available.

When defective tubing is returned to the shop from outside, and if the defect is not readily apparent, the following tests should be applied to it:

1. Test for leaks or cracks at the electrodes.
2. Test for leaks or cracks at the tubulation tip.
3. Test between letters and at all splices.
4. Test for cracks along the tubing.

The tests are listed in order of the likelihood of trouble; that is, the electrodes are most frequently found defective, leaks near the tubulation are next in occurrence, and so on. If these tests fail in disclosing the leak, the entire tubing section should be made new, since this procedure will save time in the long run.

Miscellaneous. - Many tubing manufacturers like to make tests comparing the various types of tubes, electrodes, etc. A quick force test which is commonly used and which will give a fairly good comparison is as follows. Use a tube 1 foot long with electrodes to be tested and fill only to 2 millimeters of neon. Connect to a 60-milliampere trans 1 former. Usually tubes made with this low gas pressure and connected to a 60-milliampere transformer will last only approximately 10 to 50 hours. The difference in life of various tubes can be used as being directly proportional to the life of the tubes when filled with higher gas pressure.

Whenever a hanging transformer is installed in a show window, care should be taken to shield it from the direct glare of any floodlights which may be used. If it is installed in the direct rays of such a light, it will operate at a much higher than rated temperature. A burnout may result; usually, also, the pitch filling of the transformer will begin to run out of the casing. If this happens, valuable goods on display may be ruined.

At normal room temperature, a transformer will handle a longer length of mercury tubing than it will neon. For this reason it is sometimes found that transformer manufacturers recommend a longer maximum length of blue tubing than the conditions actually warrant. If the sign is to be subjected to cold weather, the transformer may become badly overloaded as the temperature decreases and the brilliance of the discharge falls off. The current through the tube should always be kept up to the operating current.

If an especially long tube is to be bombarded, B-10 or No. 50 carrier gas or neon may be used for bombarding gas. The B-10 or No. 50 or neon must of course be pumped out after the bombarding is finished. The same procedure is followed as is given earlier.

It is often found that tubes made according to best practice do not stand up for long life. In such cases, it

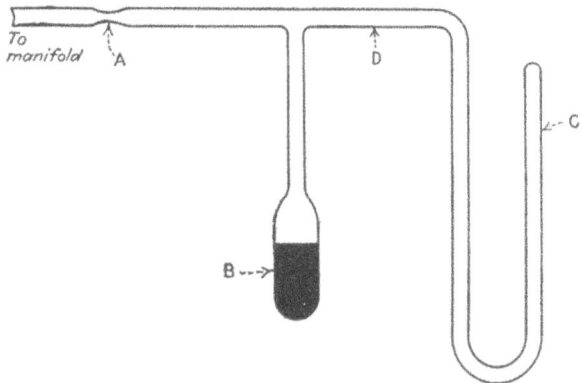

Fig. 102.—Method of constructing a mercury U gauge. The tube is pumped out thoroughly, sealed off at A, and the mercury in B poured into the U section until it fills the arm C. The gauge is then cut off and sealed to the manifold in the usual manner, at D. Tube and mercury should be heated with a soft flame before pumping.

may be that the vacuum gauge or manometer used for measuring the gas pressures is inaccurate. If the pressure is actually lower than indicated, the life of the tubes will be very short.

To test for gas on the closed arm of the gauge, allow air to enter the manifold and examine the closed leg carefully. If the mercury in the closed leg does not fill the tube completely to the top, gas is present. If it does apparently fill the tube, there still may be a small amount of gas, nevertheless, since this small amount may be compressed to a very small volume. If a McLeod gauge is available, pump the manifold out to the very best vacuum of which the pumps are capable. Measure the vacuum with the McLeod gauge. If it is below 50 microns, the level of the mercury in the two arms of the U gauge should be the same. If the mercury in the closed leg is only a small fraction of an inch lower than that in the open leg, the gauge contains gas.

If the gauge is found to be inaccurate, it should be repaired by emptying it of mercury, pumping out to a very high vacuum (5 microns or better) heating, and then filling with clean mercury, according to the procedure given in Fig. 102. After the gauge is installed on the manifold again, it should be tested by the McLeod gauge as before. If a McLeod gauge is not available, the spark coil can be used to gauge the vacuum, although it is not so reliable.

If inferior grades of rare gas are purchased, it may be found that they contain 1 or 2 per cent of helium. Even this small amount of helium will so raise the resistance of the gas that red or blue tubes filled with it will require a higher voltage per foot than usual, although the helium color will not be visible. Gas sold by reliable manufacturers and supply houses is rated on a potential gradient basis and is thus known to have the standard resistance per foot.

Black glass used for crossovers has the same disad-

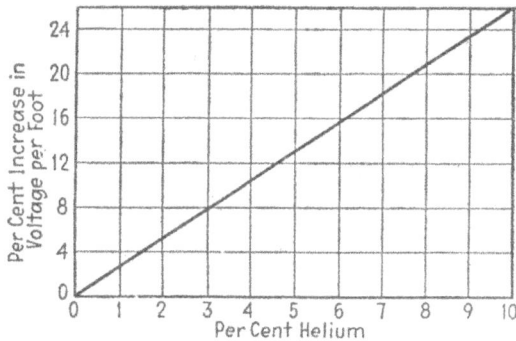

FIG. 103.—Effect of helium content on required voltage per foot.

vantage as have red and opal glass; that is, it requires a long aging period. The very highest grade of work, however, is usually done with black-glass crossovers; and if the pumping and bombarding are done properly, the aging period should not be excessively long.

Low-intensity helium gold tubes can be run from 30milliampere transformers and, if coated electrodes are used, will give long and satisfactory life when properly pumped and bombarded. If these tubes are filled to standard 11-millimeter pressure, they will run comparatively cool, though hotter than red or blue tubing. High-intensity gold tubing is much more striking and is usually desired by the customer. A low-pressure (2- to 3-millimeters) fill and a 60-milliampere transformer should be used in this case. For long life, special care must be taken both in choosing the right electrode for the job and in bombarding. A high-intensity helium tube, properly pumped and bombard-

165

ed, filled to 3-millimeter pressure, and provided with the proper electrodes will have a life equal to a 10-millimeter pressure neon tube using standard electrodes.

Many manufacturers use what is known as a pumping unit, that is, a complete bench, pump, bombarder, and glass-bending fire in one unit. Usually these pumping units are found to be too small for work or a production basis; if this happens, the fires should be removed to another bench, and the unit reserved entirely for pumping and bombarding.

China-marking crayon is usually preferable to chalk for marking glass. Chalk may eat into the glass when heated to a high temperature, whereas the china crayon will burn off in the fires.

The crossover paint can be provided with a very convenient spout, which is soldered over a hole drilled in the top of the can and provided with a cork. When the paint is used, a small amount is poured from the can into a dish, just enough to do the job. The fast-drying enamel is thus protected when not in use.

Corrosion of Lead-in Wires. - Occasionally signs must be installed in locations which are subject to more than the usual weather and atmospheric conditions. For example, signs are often installed near the sea, where salt spray is plentiful, or near factories, where the chemicals in the air may have a corrosive effect.

Whenever these conditions are encountered, special precautions must be taken to protect all metal parts of the sign, especially the lead-in wires at the electrodes. It has been found that a liberal coating of asphalt paint, applied to all exposed metal parts, both inside the sign box and out, will protect these parts against corrosion.

Blue or green tubing which has been removed from a sign for repair must be handled carefully if it is to be used again. Unless precautions are taken, tubing repumped from old mercury tubes will have a very short life. New electrodes should always be used, and unless the glass can be made perfectly clean by a nitric acid wash, it should not be used again.

Two Methods of Painting A Sign. - There are two methods of painting metal signs used for neon work. A sign can be painted either by a brush, hand applied, or by a spray gun. With a spray gun, an air compressor is required for spraying.

1. *Brushing:*

a. Wash metal with vinegar, and dry. This takes off all acid spots.

b. Coat with a metal primer which has been thinned with turpentine.

c. Apply a white lead paint, which has been first broken with a rubbing varnish. The varnish provides a binder for the white lead. The mixture of lead and varnish should be then thinned with a mixture of boiled linseed oil and turpentine until the mixture flows freely. In some cases, where a blue background is required, add a little lampblack in oil to the mixture above until it becomes gray.

d. A Japan paint of the color desired is then applied. If one coat of paint does not cover, apply another. Japan paints are to be thinned only with turpentine.

e. Use Japan color paints desired for making letters, strippings on letters, ornaments, imprints, borders, etc.

f. After the paint has dried, apply a coat of varnish.

2. *Spraying.* - When spraying a sign, at least 40 to 50 pounds' pressure is necessary. The sign should always be sprayed in a warm place having a temperature of at least 60 degrees. The following is one procedure used for spraying signs.

a. Rub down metal with pumice stone and steel wool until smooth mirror-like finish is obtained. Be sure that all foreign particles are removed.

b. Spray with one or two coats of prime lacquer.

c. Spray on lacquer used for the background of letters, border, ornaments, etc.

d. The desired shape and form of the letters and borders are then lettered in with a mask paste.

e. Spray entire sign with one or two coats, maybe three if necessary, of the paint required for the background.

f. When dry, wash sign with warm water. This washes off the mask paste, and the desired letters and borders stand out. Use only warm water in washing and stripping off sign.

g. If fancy hand designs are required, letter in all detail work.

h. When dry, lacquer entire sign with a clear lacquer.

Window Outline Borders. - Tubing is used to a considerable extent to outline store windows. The types of tubing used are as follows.

1. Outline of the window, both sides, top and bottom, requires a straight tubing or fancy tubing or tubing with fancy corner designs. In this case, the transformer is usually a hanging transformer type or a standard transformer in a metal box mounted either on top of the window or on the bottom under the window sill. In either case, tubing must be divided to bring the two leads either up or down to make a continuous circuit.

2. Window borders are also made having three sections, the top and two sides, as, when a bottom piece is used, it is subject to breakage. In this case, the tubing is brought down on either side to the porcelain housing, which is inclosed in a metal box. The terminals of the porcelain housing are then brought down through the window sill to a transformer mounted underneath.

3. In some cases, where a double border is required, this can be done by using the same arrangement as in the previous paragraph; only a double housing is put into a metal box instead of a single housing on each side. The tubing can then be connected to a transformer mounted on top instead of on the bottom. Fancy designs can be worked out using this arrangement.

4. In a great many cases, a window border is used in combination with a skeleton sign. The electrodes of the window border and the skeleton sign

must be then worked out so that all tubing is connected in series. This is especially true if the linear length of the tubing used in the combination is short enough to go on one transformer. Otherwise, two transformers must be used.

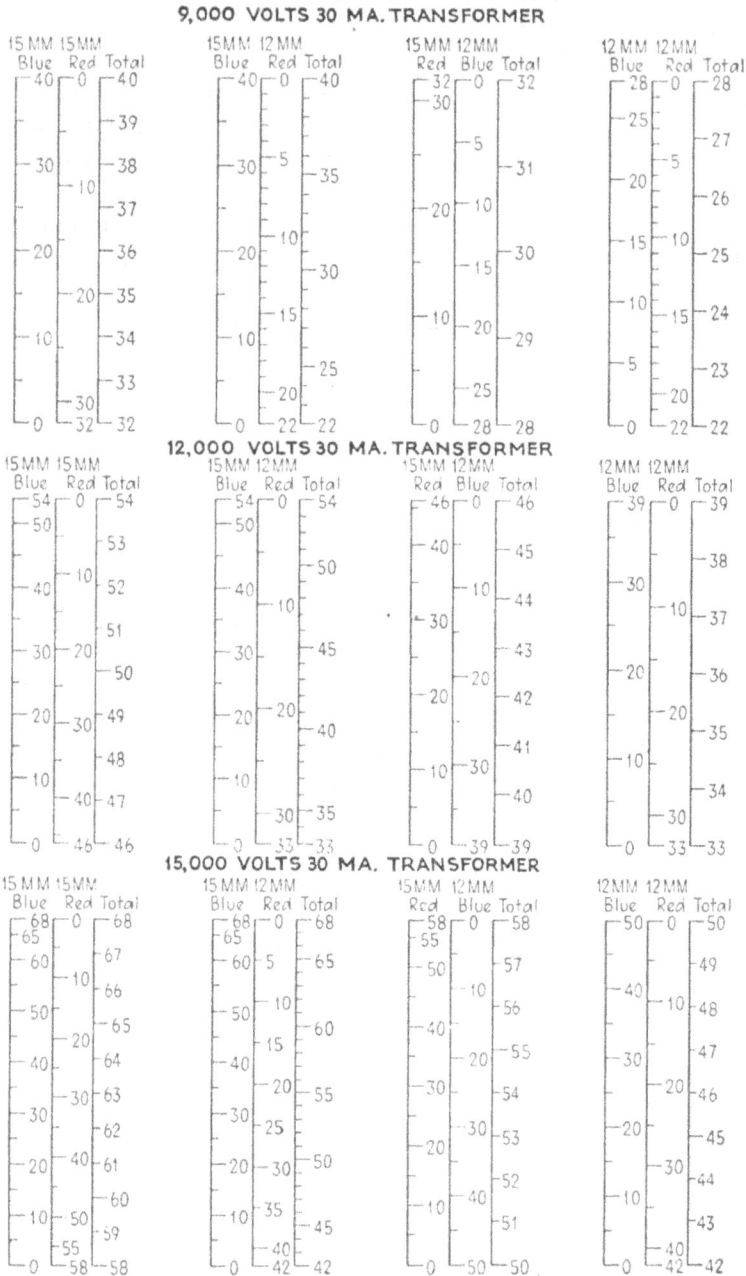

9,000 VOLTS 30 MA. TRANSFORMER

12,000 VOLTS 30 MA. TRANSFORMER

15,000 VOLTS 30 MA. TRANSFORMER

Fig. 104.—Chart for determining the combinations of red and mercury tubing which may be run from a single transformer. The use of the chart is explained in the text on the opposite page.

Two-color Sign-footage Charts. - Many signs contain more than one color; red and green, or red and blue, or red, green, and blue may often be used in a single sign. Since all transformer charts show the allowable footage of tubes of various diameters either for red or for mercury tubing, it has been difficult to determine accurately just how much tubing a single transformer will carry, if *both* red and blue tubing are used in series.

The charts given in Fig. 104 can be used to determine the proper transformer to use when two colors are to be operated from a single transformer. For example, if the sign has 30 feet of 12-millimeter red tubing for its letters, and 19 feet of 15-millimeter blue tubing for the border, look in the column for 12 millimeters red and 15 millimeters blue (the second column in Fig. 104). It will be seen that a 15,000-volt 30-milliampere transformer will carry this load. Other possible combinations may be found by selecting the numbers which lie directly opposite one another in the two columns. The total footage is given in the last row to the right in each column. In the example given above, it will be seen that opposite the figure 30 in the 12-millimeter red column lies the figure 19 in the 15-millimeter blue column (for the 15,000-volt transformer). If the combined footage is greater than indicated in any of the "total" columns, then more than one transformer must be used, and each separate color can then be operated from a separate transformer.

Flashing Diagrams. - On Fig. 105 is given a group of typical connection diagrams for obtaining different animation effects. The diagrams are self-explanatory.

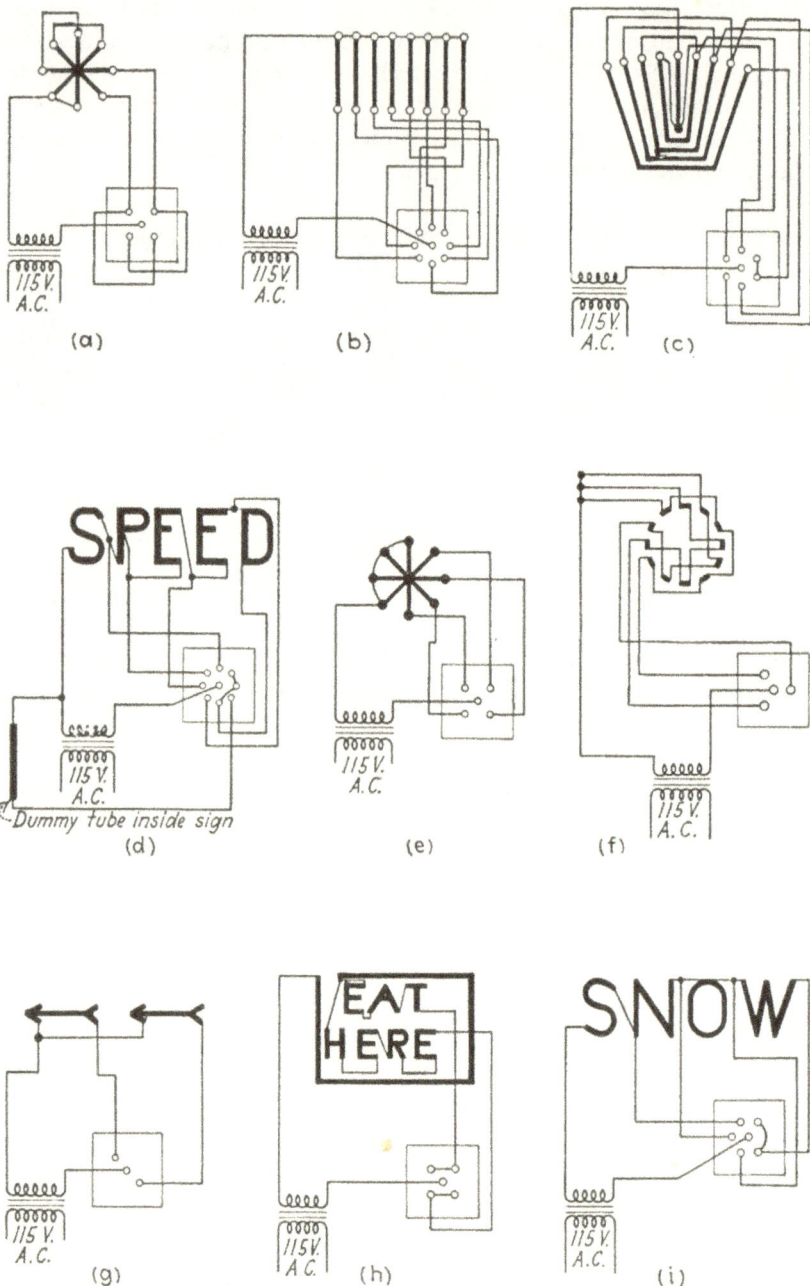

FIG. 105.—Flasher diagrams for special animation effects: (a) Alternate flashing of spokes (high speed) (b) Running effect for borders, arrows, etc. (c) Opening of fan. (d) Speller, with all-on, and all-off effect. (e) Revolving wheel. (f) Revolving circle. (g) Shooting arrows. (h) Alternate flasher with steady-burning border. (i) Speller, with all-on effect. The number of switch terminals required for each type of switching is shown in each diagram.

Appendix I - Underwriters' Laboratories' Requirements for Electric Signs

This is the requirements section of the March, 1930, edition of Underwriters' Laboratories Standard for Electric Signs.

General. - 15. This Standard covers incandescent lamp- and gas-tube signs to be employed in accordance with the rules of the National Electrical Code.

16. This Standard states minimum requirements. It is based upon records of tests and field experience and is subject to revision as further experience and investigation may show to be necessary or desirable.

17. Devices or products which comply with this Standard will not necessarily be acceptable if they have other features which, when examined and tested, are found to impair the result contemplated by this Standard.

18. A device or product having materials or forms of construction differing from those detailed in this Standard may be examined and tested according to the intent of the Standard and if found to be substantially equivalent may be given recognition.

19. Requirements are shown in distinctive boldfaced type and are supplemented by explanatory notes and descriptions of test apparatus and methods in a different type.

Sign Bodies. 20. With the exception of wood employed for the external decoration of signs and kept at least 2 inches from the nearest lamp holder, electrode receptacle, or gas tube, sign enclosures shall be constructed entirely of metal or other non-combustible material. Materials other than sheet metal shall be made the subject of special investigation before being used.

21. Sheet steel faces, sides, tops, and bottoms of signs shall be galvanized when the metal is lighter than No. 20 U. S. sheet metal gauge. Such parts made of sheet steel No. 20 U. S. sheet metal gauge or heavier or made of metals other than steel shall be galvanized, treated with not less than three coats of paint, or otherwise acceptably protected against corrosion.

22. Vitreous porcelain enamel may be used for the protective coating on sheet steel of No. 20 U. S. sheet metal gauge or heavier.

23. Sheet metal faces, sides, tops, and bottoms shall be not lighter than specified in the following table:

24.

	Gas tube or combination gas tube and incandescent lamp signs		Incandescent lamp signs
	Flat sheets	Crimped sheets[1]	Flat sheets
Sheet steel.......	24 U. S. gauge	26 U. S. gauge	28 U. S. gauge
Sheet copper......	20 oz. (0.028 in.)	....	16 oz. (0.022 in.)

25. [1] A sheet to be considered crimped shall be ribbed, corrugated, or embossed over its entire surface.

26. The thickness of sheet metals other than steel or copper is not specified; but such metals will be generally acceptable if the thickness is not less than that specified for sheet copper.

27. Raised or channel metal letters to which tubes are attached shall be not lighter than No. 28 U. S. gauge. When any dimension of the letter exceeds 15 inches, the metal shall be not lighter than No. 26 U. S. gauge.

28. An individual metal sheet will be considered acceptable for the gauge in question if the average thickness as measured in accordance with paragraph 30 is not more than 10 per cent under the thickness for that gauge as given in the following table:

29.

U. S. gauge	Minimum average thickness, in inches	
	Uncoated	Galvanized
28	0.0138	0.0176
26	0.0166	0.0204
24	0.0221	0.0259
22	0.0275	0.0314
20	0.0331	0.0370

30. The average thickness of a sheet shall be determined by five micrometer readings spaced equally across the width edge of the full sheet as rolled. (It will usually be found that the short edge of full sheets is the width edge as rolled.) Readings shall be taken at least an inch in from edges of the sheet.

31. The sign body shall be so constructed as to secure ample strength and rigidity.

32. Acceptable rigidity can be obtained by reinforcing the sign with partitions, channels, angles, and strap iron braces. Such reinforcements are usually found necessary for all gas tube signs because of the added weight of transformers within the signs and the fragileness of the tubes supported by the faces. In the case of large signs, acceptable strength and rigidity can be obtained through the use of a structural iron framework to which reinforced faces can be secured.

33. Sheet metal faces secured to sheet metal or structural channel frames, and seams in sheet metal faces or at edges shall be welded or secured by rivets or bolts spaced at not over 8-inch intervals, except that when the faces are made of No. 20 U. S. gauge or heavier metal or when the seam is made so that three thicknesses of metal are engaged by the rivets or bolts, the spacing may be increased to 12-inch intervals.

34. Where sheet metal screws are used in sign construction in place of rivets or bolts, they shall be so located or protected that the points of the screws will be unable to cause injury to the insulation of any of the conductors in either the primary or the secondary wiring.

35. It is strongly recommended that the use of sheet metal screws in sign construction be limited to locations where it is particularly difficult or infeasible to use rivets or bolts.

36. NOTE. - The acceptance of sheet metal screws is on a tentative basis. If field experience with this method of sign construction is not favorable, the preceding paragraphs will be deleted from these requirements.

37. Faces with a length or width not in excess of 10 inches, or when of copper, need not be secured as specified above, provided that the joints are continuously soldered their entire lengths.

38. Sheet metal channels or strap iron used to reinforce faces shall not depend solely upon solder for securing them in place.

39. The body assembly, if for outdoor use, shall, in so far as is practicable, provide a weatherproof enclosure. Doors shall effectively exclude rain from entering the enclosure.

40. The requirement of the preceding paragraph will necessitate the soldering of seams on faces unless they are formed so as to shed water. When a gas tube passes through a wall, it is recommended that a barrier of insulating material be used to close the opening, and such a barrier shall be required when the tube passes through the top of a sign.

41. Doors or covers on signs wired throughout with uninsulated conductors shall give access to transformer compartments only.

42. The requirement of the preceding paragraph necessitates the provision of means for the renewal of tubes from the exterior of the sign enclosure and the isolation of the transformer in a transformer compartment.

43. Each compartment of a sign intended for outdoor use shall be provided with at least one hole not less than ¼ inch nor more than ½ inch in diameter to permit drainage of moisture or condensation.

44. All portions of structural iron or steel parts shall be galvanized, enameled, or be provided with an equivalent protection against corrosion.

Sign Supports. - 45. Each sign shall be provided with means for attaching it to a building wall or to a support or hanging rig.

46. It is impractical to outline methods for the supporting and staying of signs which would be sufficient in all instances. In general, the method shall be such as to give a reasonable factor of safety above the strains incident, not only to the weight of the sign but to wind pressure and other strains to which they may be subjected in service. Individual letters and displays shall be provided with means for attaching them to a background or framework (building wall, structural framework, wire mesh framework, etc.); and when not of bolted, riveted, or welded construction, they shall be arranged for attachment to the background or framework at not over 2-foot intervals (vertical and horizontal).

Glass Panels and Letters. - 47. Glass panels may be used in combination gas tube and incandescent lamp signs provided breakage or removal of the panels will not give access to any part of the high voltage circuit.

48. Panels or letters not included in these requirements shall be made the subject of special investigation with respect to strength, rigidity, and security of fastening before being used.

Thickness of Glass. - 49. The thickness of glass used for panels and individually moulded letters in electric signs shall be not less than given in the following table:

50.

Maximum size of glass panels		Maximum dimension of individually moulded letters, inches	Minimum thickness of glass, inches
Dimensions, inches	Area, square inches		
20	250	20	$\frac{1}{8}$
30	500	30	$\frac{5}{32}$
60	900	. . . . . .	$\frac{1}{4}$
Over 60	Over 900	Over 30	Special investigation

51. The area of an individual glass panel shall not exceed 25 square feet (3600 square inches).

52. Wired glass shall be used for panels exceeding 900 square inches in area unless the mounting is such that the possibility of breakage of the glass due to warping or bending of the supporting face or surface is entirely eliminated.

53. Wired glass is recommended for all panels having areas in excess of 500 square inches.

Fastening for Glass Panels and Letters. - 54. Clips or troughs for securing glass panels or letters in signs shall overlap the glass not less than ½ inch, shall be not less than ½ inch in width, and shall be of a thickness not less than that given in the following table:

55.

Maximum area of panels, square inches	Maximum dimension of moulded letters, inches	Thickness, in inches			
		Galvanized steel	Phosphor bronze	Copper	Terne plate
500	30	0.0188	0.020	0.022	0.013
Over 500	Over 30	0.0380	0.045	0.045	

56. When terne-plate clips are used there shall be at least one clip on each glass edge for every 3 inches of glass or fraction thereof, and clips shall be spaced not more than 1½ inches from corners. Except as noted in the following paragraph, when steel, phosphor bronze, or copper clips are used, there shall be at least one clip on each edge for every 6 inches of glass or fraction thereof, and clips shall be spaced not more than 3 inches nor less than 1½ inches from corners.

174

57. For raised glass letters which are designed to fit closely into openings cut in the sheet metal face, the spacing between clips may be increased so that one clip is provided for every 8 inches of glass edge or fraction thereof; clips may be omitted on side edges which do not exceed 8 inches in length, and one clip may be used on a side edge which does not exceed 10 inches in length; and one clip may be used on a top or bottom edge which does not exceed 12 inches in length, where each side edge is provided with a clip for every 6 inches of glass edge or fraction thereof.

58. Horizontal troughs for supporting glass panels or letters shall be provided with drain holes if such troughs are so arranged, formed, or located that water pockets may be formed.

Supports for Glass Panels or Letters. - 59. Sheet metal surfaces or faces of electric signs supporting glass panels or letters shall be of sufficient thickness to give mechanical strength such that the glass will not be subject to breakage due to bending or warping of the metal; or such metal shall be reinforced by means of suitable strap iron or angle iron or sheet metal channels to give the necessary rigidity.

60. Individual sheet metal sections supporting glass letters or panels, as used in signs with interchangeable letters, shall be supported at top and bottom edges by troughs overlapping the edges not less than ½ inch. Each sheet metal supporting section shall have an offset not less than ½ inch wide extending the length of one vertical edge so as to form a closely fitting overlap with the adjacent section.

61. Special features involving the use of glass in electric signs and not covered by these requirements shall be made the subject of special investigation before being used.

62. A combination gas tube and incandescent lamp sign shall be so constructed that uninsulated high voltage parts will not be exposed to contact by persons relamping the sign.

63. In the preceding paragraph, "relamping" applies to the replacement of incandescent lamps.

Wiring

Low Voltage (600 Volts or Less). - 64. All wiring shall be done with type-R rubber-covered wire, and when six or more conductors are bunched together, or grouped in a trough or gutter, they shall be of "Flame-retarding" type-R wire having the outer braid treated with a flame-retarding as well as a moisture proofing compound. [1]

65. The following maximum allowable current-carrying capacity of copper wires shall not be exceeded; and wires shall be not less than No. 14 B. & S. gauge in size, except that the flexible cord and other low voltage wiring of a portable sign for indoor use may be No. 16 or No. 18 B. & S. gauge.

66.

B. & S. gauge	Amperes	B. & S. gauge	Amperes
18	3	8	35
16	6	6	50
14	15	5	55
12	20	4	70
10	25		

67. Wires shall be neatly run and fastened so as to be mechanically secure.

68. Wires passing through partitions shall pass through smooth, non-combustible bushings firmly fixed in place.

69. The bushing may be of metal or of a non-combustible insulating material. Short lengths of standard flexible non-metallic tubing or porcelain tubes may be used in place of bushings to protect the wires. If such tubing or tubes are used, they shall be securely fastened to prevent them from moving along the wire.

69a (formerly paragraph 103). Wireways within sign enclosures should be formed of metal not lighter than No. 24 U. S. gauge.

70. Lamp holders shall be so connected that each circuit operating at 125 volts or less may be protected by a fuse of rated capacity not greater than 15 amperes and that each circuit operating at 126 to 250 volts may be protected by a fuse of rated capacity not greater than 10 amperes.

71. Circuit wires connected to the screw-shell terminals of lamp holders shall be identified by means of a white or natural gray braid. The braid of all other wires in the sign shall be of a contrasting color.

72. Each sign of the gas tube type shall be so wired that not more than one transformer will be dependent upon a single automatic overload protective device unless the combined load is less than 1650 voltamperes.

73. The above is interpreted as requiring a separate circuit for each transformer or a separate circuit for a group of transformers which have a combined load of less than 1650 voltamperes. Thus, four 400 volt-ampere transformers may be grouped on a single circuit, whereas five 400 volt-ampere transformers necessitate at least two circuits.

Connections and Splices. - 74. Wires shall be so spliced or joined as to be both mechanically and electrically secure and shall, in addition, be soldered. The joints shall then be covered with both rubber and friction tape to a thickness at least equivalent in insulating value to the insulation of the joined conductors.

75. Wires shall be fastened to lamp holder terminals so as to be mechanically and electrically secure, and shall, in addition, be soldered.

76. The insulation on the wires at terminals shall not be stripped back so far that the bare wire will overhang the edges of lamp holder bases.

77. Terminals at lamp holders and exposed parts of wires at the lamp holder terminals shall be treated to prevent corrosion.

78. A standard insulating compound or paint may be used to prevent corrosion at lamp holder terminals.

79. When a combination or gas tube type sign built in sections or involving a number of individual letters is supplied with low-voltage lead wires which must be connected up after installation, a sheet metal box or enclosure not lighter in gauge than the metal of the sign itself shall be provided in each section or letter to house the splices, unless the arrangement and location of the conductors is such that there will be no reduction of the spacing between high and low voltage wires as required elsewhere in this standard.

High Voltage (Above 600 Volts). - 80. Except as noted in the following paragraph, the high voltage wiring of each sign shall be complete before it leaves the factory.

81. The high voltage wiring of a sign, the transformer for which is to be located outside the sign enclosure, need not be complete before it leaves the factory, if all necessary terminals, bushings, fittings, etc., are in place as part of the sign and are so arranged that the necessary connections can be readily made upon installation.

82. Insulated conductors shall be standard, suitable for the voltage involved, not smaller than No. 14 B. & S. gauge, and of the stranded type.

83. Uninsulated conductors (as used in signs wired throughout with uninsulated conductors) shall be not less than No. 10 B. & S. gauge.

84. Conductors having circular-mil areas less than 3800 will not be considered as No. 14 B. & S. gauge.

85. It is recommended that insulated conductors be used and that all terminals, connections, splices, and other live parts be insulated.

86. Signs wired with uninsulated high voltage conductors shall be made the subject of special investigation before being used.

87. The conductors shall be supported as close as practicable to the terminal connections, both at transformers and electrodes.

88. When no support is provided at an electrode terminal, it will usually be necessary to provide a cable support within 6 inches of the electrode connection.

89. It is strongly recommended that supports be spaced not more than 18 inches apart.

90. Except at transformer connections, insulated conductors shall not be spliced. At transformers or electrodes, connections shall be mechanically secure. Stranded conductors shall be soldered together before being fastened under clamps or binding screws.

91. The use of a grommet or its equivalent is recommended where the conductor is secured at a binding screw or post.

92. Where the high tension wiring is connected to a transformer of the weatherproof type having secondary leads attached, joints shall be made mechanically secure, shall be soldered, shall be covered with rubber tape to give an insulation equivalent to that on the conductors and shall have an over-all winding of friction tape.

93. At gas tube electrodes, connections between high tension conductors and electrode leads shall be soldered or made with standard wire connectors suitable for this particular use.

94. Bushings or fittings used where secondary conductors emerge from the sign enclosure shall be of the weatherproof type or shall be so located as to prevent rain from entering the enclosure.

95. There shall be no condensers nor current interrupting devices in the high voltage circuit.

96. Uninsulated high voltage parts of a portable sign or a sign intended for indoor use shall be wholly inaccessible or the enclosure shall be provided with an interlock switch or switches arranged so that the opening of any door or cover giving access to such high voltage parts will positively cause the breaking of the primary circuit.

97. The requirement of the preceding paragraph will necessitate the provision of the interlock arrangement for a portable or indoor type of sign in which access to the inside of the enclosure is necessary in order to replace tubes; unless the construction is such that there will be no possibility of exposure of live parts to accidental contact when tubes are being replaced.

98. A portable sign or an indoor sign which is intended to have the tubes replaceable without opening the enclosure shall be so constructed that no live parts or live connections will be exposed to accidental contact when tubes are being replaced.

99. A portable sign or an indoor sign shall be wired throughout with insulated conductors suitable for the voltage involved.

100. A portable sign shall have the transformer and all wiring except the flexible cord to the source of supply housed within the enclosure.

Spacings. - 101. The spacings between high voltage parts in an electric sign as described below shall be not less than those given in the indicated column of the following table for the voltage corresponding to the open circuit potential of the transformer employed.

A

101 A. Between uninsulated live parts of opposite polarity, whether the mid-point of the transformer secondary is grounded or not.

102. Between uninsulated live parts and grounded metal, when the mid-point of the transformer secondary is not grounded.

102A. Between uninsulated live parts of different secondary circuits (where two or more transformers are used), whether the mid-points of any or all of the transformer secondaries are grounded or not. Where two or more transformers are employed and where different secondary potentials are involved, spacings are to be based upon the highest open circuit transformer voltage.

B

103. Between uninsulated live parts and insulated primary conductors, when the mid-point of the transformer secondary is not grounded.

103 A. Between uninsulated live parts and grounded metal, when the mid-point of the transformer secondary is grounded.

C

104. Between uninsulated live parts and insulated primary conductors, when the mid-point of the transformer secondary is grounded.

104A. Between insulated conductors of opposite polarity in all cases.

105. Between insulated conductors and grounded metal in all cases.

106. Between insulated conductors of different secondary circuits in all cases.

107. Between insulated conductors of primary and secondary circuits in all cases. [2]

108.

Transformer secondary rating, in volts	Minimum spacing through air or over surface, in inches		
	A	B	C
5,000	1	¾	½
10,000	1½	1	¾
15,000	2	1½	1

Barriers. - 109. Barriers introduced to effect spacings shall be of standard, non-combustible, non-absorptive insulating material.

110. Insulating tubes on high voltage conductors to effect spacings shall be taped in position or otherwise acceptably secured so as to prevent their sliding along the conductors.

Transformers. - 111. Standard gas tube sign transformers shall be used.

112. Transformers shall be secured in place in a reliable manner but shall not be mounted on doors or hung so that the entire weight is suspended from a single sheet metal side, face, or top.

113. It is recommended that transformers be supported by the angle-iron framework rather than the sheet metal of the sign. However, strap-iron brackets or sheet metal channels, not lighter than No. 24 U. S. gauge when secured by at least two rivets or bolts to each face of the sign, may be used as supports for transformers. It is further recommended that transformers be supported in a horizontal position and right side up.

114. Transformers unless of the weatherproof type or unless of the type having sealed-in windings, if on or within the body or structure of the sign, shall be in a separate accessible weatherproof enclosure of metal of thickness not less than No. 28 U. S. gauge.

NOTE. - 115. Installation rules of the National Electrical Code require that transformers if otherwise located shall be enclosed in standard cut-out boxes or cabinets.

Lamp Holders. - 116. Lamp holders for sign use shall be of the keyless, porcelain or moulded composition type. Standard lamp holders shall be used.

117. Miniature lamp holders are not considered acceptable for use in outdoor signs.

118. Lamp holders shall be so installed as to afford permanent and reliable means to prevent possible turning.

119. The more common methods of securing lamp holders to sign faces to prevent them from turning relative to the sign faces are: two screws; a lug on the holder engaging a notch or indentation at the hole in the sign face; a notch in the holder engaging a tongue or the equivalent at the hole in the sign face. The mere fictional contact of screw rings or gaskets, etc., with sign faces will not be acceptable.

120. Lamp holders shall be so placed that terminals will be at least ½ inch from terminals of other lamp holders and from the metal of the sign, or an equivalent separation shall be obtained by means of barriers of standard insulating materials.

Cable-and-tube Mounting. - 121. Supports for high voltage conductors shall be of non-combustible, non-absorptive insulating material and, if of porcelain, shall be glazed on exposed surfaces.

122. Fiber, rubber, hot moulded shellac, or phenolic compositions are not considered acceptable material for cable supports.

123. Each cable support shall provide a reliable means for mounting the high voltage conductor, shall have provision for being securely attached to the sign body, and shall be such that the spacings called for elsewhere in these requirements will be maintained.

124. Supports for glass tubing shall be of the insulated type, of non-absorptive, non-combustible material, and preferably adjustable.

125. Fiber, rubber, and hot moulded shellac compositions are not considered acceptable materials for tube supports.

126. The supports shall provide a reliable means for mounting the tubing, shall have provision for being securely attached to the sign face, and shall maintain the tubing not less than ¼ in. from grounded metal parts. This spacing shall be maintained where the tubing passes through the sign face, if of metal.

127. It is recommended that bushings of non-combustible insulating material be used where the tubing passes through the metal faces of the signs.

128. It is strongly recommended that glass tubing be not supported on doors or covers of signs.

Electrode Receptacles. - 129. It is strongly recommended that electrode receptacles be used.

130. Electrode receptacles for gas tubes shall, if used, be of a non-combustible, non-absorptive insulating material.

131. Fiber, rubber, hot moulded shellac, or phenolic compositions are not considered suitable materials for electrode receptacles or for the mounting of high potential parts.

132. A rubber gasket may be used in connection with the mounting of an electrode receptacle.

133. A rubber gasket may be used around the gas tube in an electrode holder to prevent the entrance of dust, moisture, etc., provided that the gasket is not depended upon for the insulation of the tubing and is not in contact with grounded conducting material.

134. An electrode receptacle shall not permit water to accumulate therein and thereby form a conducting path to grounded metal such as the sign face.

135. Each electrode receptacle shall be of substantial construction and shall provide means for reliably securing it to the sign face. Electrode receptacles shall not be secured to sign faces by means of sheet metal screws.

136. Each electrode receptacle shall be constructed so that, when installed, the spacings mentioned elsewhere in these requirements will be maintained.

137. It is strongly recommended that the design of an electrode receptacle be such that when installed with all conductors connected, there will be no exposed live metal parts.

Dielectric Strength. - 138. Accessories such as bushings, cable supports, electrode receptacles, etc., used in signs in which the high voltage conductors are not provided with insulation, shall be capable of withstanding an alternating-current potential of 45,000 volts applied for one minute and as measured by sphere gaps. Bushings shall be tested for surface creepage and for puncture resistivity.

Capacitors. - 139. A capacitor intended for power-factor correction shall be capable of withstanding without breakdown for a period of 15 minutes a direct-current potential of five times its rated voltage.

Connection to Supply. - 140. A sign for permanent installation shall be provided with means for connection to conduit, armored cable, or open wiring. Provision for connection to open wiring shall be so located or arranged as to prevent water entering the sign.

141. The above requirement will have been complied with if, for conduit or armored-cable connection, a standard conduit box is rigidly secured to the sign or a chase nipple is held in place by a locknut. and. if for open wiring, an insulating bushing is secured in place.

142. It is recommended that provision for the supply connection be located elsewhere than on the top of the sign.

143. A portable sign shall be supplied with a length of standard flexible cord and attachment plug for connection to the source of supply. A suitable strain relief shall be provided in the cord; and except for jacketed cord such as types P, S, or SJ, the cord shall enter the enclosure through a noncombustible insulating bushing.

144. Type PO-64 cord will not be considered acceptable as the flexible cord for a portable sign as referred to in the preceding paragraph.

Cut-out Bases, Panel Boards, Etc. - 145. Standard cut-out bases, panel boards, flashers, switches, and other similar devices shall be used when they are provided as part of the sign assembly. If on or within the body or structure of the sign, they shall be in a separate, accessible, weatherproof enclosure of metal or thickness not less than No. 28 U. S. gauge.

NOTE. - 146. Installation rules of the National Electrical Code require that the above devices if otherwise located shall be enclosed in standard cut-out boxes or cabinets.

Marking. - 147. Each sign shall be marked with the manufacturer's name where it will be readily visible after installation.

148. Each sign of the gas tube type or combination incandescent lamp and gas tube type shall be marked with the input in amperes (volt-amperes or kilovolt-amperes) and volts so as to be readily visible after installation. In the case of the combination type, it shall be made clear that the rating applies to the tubes and not to the incandescent lamps.

[1] Revised August, 1932.
[2] Paragraphs 101 to 108, inclusive, revised August, 1932.

Appendix II - Typical One-Man Plant

A Complete List of All Requirements

This combination of machinery, equipment, and material was decided upon as the ideal arrangement for a one-man plant after studying the needs of new sign companies. With the amount of raw materials listed here, neon signs can be manufactured as soon as the machinery is in operation.

1 5-way crossfires with adjustable sliding bases and special mixers.

1 small tipping hand torch.

1 large splicing hand torch.

2 mixers for torches.

1 12-inch improved "fishtail" ribbon burner with special needle-valve mixer. 1 blower (with motor).

1 mechanical compound oil pump (complete with motor). 1 butyl phthalate or mercury condensation pump. 1 5-kilovolt-ampere bombarding transformer.

1 regulator for above.

2 feet vacuum rubber tubing. 2 8-millimeter stopcocks.

2 3-millimeter stopcocks.

6 2-millimeter stopcocks.

1 gas-pressure gauge (butyl type).

1 liter neon gas.

1 liter No. 50 gas for blue and green tubing (or B-10 gas)

1 testing coil (sparker). 100 feet mica cable. 20 feet glass-blower's rubber tubing. 50 pounds 15-millimeter clear-glass tubing. 50 pounds 12-millimeter clear-glass tubing. 50 pounds 10-millimeter clear-glass tubing. 50 pounds 9-

millimeter clear-glass tubing. 10 pounds 5-millimeter clear-glass tubing for tubulating.

10 pounds 15-millimeter noviol tubing (for green).

10 pounds 12-millimeter noviol tubing (for green).

200 large electrodes.

100 small electrodes.

Accessories

In the preceding list are given the machinery and equipment needed in a neon plant. These are the chief items, but, before a plant can operate efficiently, a number of accessories must be assembled and on hand.

Pumping -system Accessories:

1. Switch for pump motor and condensation pump heater.

2. Switch for bombarding transformer.

3. Vacuum rubber tubing of varying widths for hand torch, pump connections, etc.

4. Lengths of high-tension cable with rubber-insulated battery clips for connecting tubing unit to bombarder.

5. Glass manifold (should be constructed at shop by glass bender).

6. Pieces of white paper to indicate temperature of glass during bombardment.

7. Acids and solvents for cleaning out dirty and stained tubes, etc.

8. Supply of especially rectified oil for vacuum pump.

9. Triple-distilled mercury, C. P. U. S. P.

10. Mica pieces for insulating overlapping parts of tube during bombardment.

11. One tin stopcock grease.

Glass-bending Accessories:

1. Asbestos millboard, or "Transite," covered worktable, 2½ feet high, broad surface.

2. Four hardened-steel glass-cutting files.

3. 1-pound assorted corks.

4. Marking chalk or crayons.

5. Conveniently located deep racks and bins for storing assorted sticks of glass tubing, electrodes, etc.

6. Racks to hold completed glass tubing.

Aging Accessories:

The following items are required to take care of the aging process:

1. Aging table with ample surface covered with asbestos millboard. Height 36 inches.

2. Switch for operating aging transformer.

3. High-tension leads, clamps, etc., for connecting tubing to aging transformer.

4. Aging transformer - 15,000 volts, 60 milliamperes.

Layout Accessories:

The following items constitute a few suggestions as to the principal materials required for a layout department:

1. Complete drawing table and stool.

2. Powdered charcoal and powdered chalk.

3. Heavy black marking pencils, ordinary pencils, colored crayons, chalk, etc.

4. Rulers, T square, miscellaneous drawing instruments.

5. Tags, markers, etc., for recording and classifying templates and asbestos layouts.

6. Perforating wheel.

7. One roll 36-inch heavy black carbon paper.

8. One roll heavy brown paper for templates.

9. One roll wide tracing paper.

10. One 50-pound roll of 36-inch 020 nonburn asbestos.

11. Map measuring wheel.